Installing TV & FM Antennas

Installing TV & FM Antennas

Leo G. Sands

With a specially written chapter for the guidance of the English reader by W. Oliver

FOULSHAM-TAB LIMITED
Slough Berks England

Foulsham-Tab Limited
Yeovil Road Slough Berks England

Installing TV & FM Antennas

Library of Congress Number 74-75210
Cat. Code No. 636

ISBN 0-7042-0145-3

Introduction Printed and Made in Great Britain by
A. Wheaton & Co., Exeter

Balance printed in U.S.A.

It is essential the English reader should read this chapter.

In most areas, the effect of a suitable and efficient aerial system is wholly beneficial to reception of both television and VHF/FM radio transmissions.

If you live in these areas, a first-rate antenna can do nothing but good, especially for colour TV and stereo or quadraphonic reproduction of radio programmes.

Admittedly in certain other areas there are exceptions, notably where a receiver is used virtually "under the shadow" of a powerful transmitting station or relay booster. In such cases an ambitious antenna may be quite unnecessary, and indeed the field-strength of the received signal may be such that deliberate attenuation may be needed to avoid overloading the set.

But apart from exceptional cases like this, the right choice of aerial or antenna will make all the difference to the quality and consistent reliability of your reception, provided the system is properly installed. Optimum height and siting, taking advantage of directional properties, will ensure maximum signal pick-up and suitable low-loss feeder arrangements will ensure that this is fed into the set with minimum loss on the way.

This book will help you to choose exactly the right kind of antenna for your needs and to install it yourself in the best way, thereby optimizing results and saving money at the same time.

Some parts of the book *will need re-orientating* for use in this country, because the author is discussing the subject from the American point of view, since this book is of American origin.

One purpose of this introductory chapter to the English edition is to help you make the necessary adjustments to text and illustrations to suit the scene in Britain where TV and radio are concerned.

A wide variety of aerial systems discussed and illustrated in these pages, range from the simplest dipole to quite elaborate stacked arrays of multi-element types. These are of conventional design. Less conventional and sometimes highly original types have, from time to time, been devised and described by amateur or semi-professional experimenters, writing in various technical journals.

The whole subject of TV, and VHF/FM radio, antenna systems is a fascinating one and despite the many existing designs, this field still affords scope for enterprising experiment.

In the interests of safety it is absolutely essential to take proper precautions against any risk of any aerial system becoming "live" to the mains.

There are many types of indoor and outdoor antennas which will give good results; but in every case the first consideration should be the safety aspect. Unless adequate precautions are taken to prevent any possibility of the aerial becoming "live" to high voltage, anyone who happens to touch the exposed metalwork of the aerial itself or its associated accessories may be at risk; and if a "live" aerial and any earthed object are touched simultaneously the risk could be a lethal one.

In fact, a well-known London coroner told the present writer, some years ago, that the most gruesome case of electrocution that he had ever encountered in his professional capacity resulted from a live TV aerial. The victim was orientating a directional dipole system in a loft; he was standing on an earthed metal water-tank and holding the aerial elements with both hands. When the aerial (through a fault condition) became live to the mains he was unable to let go and was electrocuted.

Another (more recent) fatality occurred through someone touching a set-top antenna and earthed electric-fire simultaneously. The antenna was originally isolated by the manufacturer, but a subsequent modification nullified this safety precaution and set up a path of conducting continuity between the metal aerial and the high-voltage mains.

The exact form the precautions should take will depend on individual circumstances, and in any case of doubt you should seek expert advice on the safety aspects before experimenting with or permanently installing any aerial, especially an indoor type which can be only too easily handled or contacted by the users of the set.

Make sure that any isolating device for aerial safety incorporated by the manufacturer of the set is still in order and has not been tampered with or eliminated by any subsequent unorthodox modifications or incompetent servicing. If fully adequate precautions have not already been taken, this must be done before experimenting with aerials having exposed metalwork.

On pp. 14–15 the author describes set-top antennas—monopole telescopic, "rabbit ears", and UHF loop—popular in the States. All of these types have been used here also; but a good many up-to-date sets in the portable category are fitted with a UHF loop antenna (a "halo" shape) only. This is because they are single-standard sets for use in areas where all three existing programmes—BBC1, BBC2 and ITV—are available on the UHF band (625 lines and FM intercarrier sound).

At the time of writing, the obsolescent VHF TV stations are still working, using 405 lines and AM sound; but they are eventually to be phased out in favour of the UHF transmissions giving better picture definition and higher fidelity sound-quality. Older, dual-standard sets are accordingly equipped with two antennas for portable use, the telescopic or rabbit ears types for VHF and the halo loop for UHF; also two separate co-ax sockets for extra aerials if desired as an alternative, these being sited outdoors, on rooftop or mast, or in a loft.

TV aerials have, for a good many years now, been grouped dimensionally (e.g., Group A, Group B, etc.) to resonate within the UHF TV bands and to suit the various channels. Colour coding is used to differentiate—Red, A; Yellow, B; Green, C.

The information on pp. 23–25 applies to American TV allocations and MUST be disregarded by British readers planning aerials etc. for TV and FM reception in the United Kingdom.

In general, all information regarding frequencies, aerial dimensions, etc., must be adjusted in relation to British channel allocations for transmitting stations, and not the American ones which despite some broad general similarity in some respects are subject to important variations.

Some of the specific details in the chapter on Transmission Lines (4) refer to American practice, components, etc. and must be modified to suit British practice in every necessary detail.

The remarks on CATV systems (p. 111 etc.) apply to American systems. Corresponding systems established in Britain serve the same purpose by similar general methods, but the practical distribution systems etc. may differ, and "terminal units" may be fitted instead of conventional receivers. In one typical system, a rotary station-selector switch fitted just inside the window selects any one of several programmes (e.g., BBC1, BBC2, ITV) and the resultant signal is fed to the terminal unit which looks, externally, like a normal TV receiver, but is simpler and cannot be used independently of the cable system.

In the chapter (10) on Reception Problems, mention is made of interference from CB and "ham" radio transmitters. Citizens Band is a radio communication facility available in the States but not here (at present).

Amateur transmitting stations (totalling thousands) are active in Britain (as in most other countries) on allocated frequency-bands which they are licensed to use. (The 6-metre band mentioned on p. 153 is not among those for which British amateurs are licensed, at present.)

The FCC (p. 154) deals with telecommunication licencing, frequency allocations, etc. in the States. Similar duties are carried out here in Britain, but by departments which have been changed from time to time. Originally these matters were handled by the Post Office; later by the Ministry of Posts and Telecommunications; and, since April 1974, by the Home Office which took over responsibility for the former MPT departments dealing with broadcasting and radio-regulatory duties.

In cases of interference caused by licenced amateur transmitting stations, the Radio Society of Great Britain (headquarters address: 35 Doughty Street, London WC1N 2AE) is often helpful in solving technical problems.

Turning to another matter: American domestic electricity supply arrangements differ from ours not only in regard to voltage and frequency (normally 110–120 VAC 60 Hz instead of our 240 VAC 50 Hz) but also in connection modes and grounding (earthing); so arrangements advocated in American technical literature *must* be checked as to suitability *and safety* for use in this country, where the earthing arrangements etc. are liable to differ. In any case of doubt, especially on safety aspects, expert technical advice should be sought locally.

Many manufacturers, mail-order and retail suppliers are currently engaged in the production and distribution of antennas and components or materials suitable for installation of complete aerial systems, indoors, outdoors or mobile.

Some useful names, with brief indication of locality, include the following (covering not only antenna specialists but general electronic suppliers handling goods relating to antenna systems):

Antiference Ltd. (Aylesbury, Bucks); Chas. H. Young (Birmingham); Colomor Electronics (London W4); Communique (London NW8); Crescent Radio (London N22); C. T. Electronics (London W4); D. P. Hobbs Ltd. (Luton); East Cornwall Electronics (Callington); Felstead Electronics (Cheadle); Garland Bros. (London SE8); Henry's Radio (London W2); Hills Components Ltd. (Watford); H. L. Smith (London W2); Home Radio (Mitcham CR4 3HD); Integrex Ltd. (Derby); J-Beam Aerials Ltd. (Northampton); J. Bull Electrical Ltd. (Croydon); Lasky's Audiotronic (London NW9); Milward's (Birmingham); Mosley Electronics (Norwich); Newmart Electronics (Manchester); Partridge Electronics (Broadstairs); Sounds Systems of Suffolk (Ipswich); South Midland Communications (Southampton); Teleservice (Tunbridge Wells); Trampus Electronics (Windsor); and Yates Electronics (Bedford).

For full addresses, telephone numbers, terms of business, catalogue charges, etc., see these firms' current advertisements in latest issues of technical journals. Some mail-order and/or retail firms have only one address, but others (such as Lasky's) now have a number of branches in various parts of the country; there may be one near enough to your own locality to make personal shopping feasible.

Preface

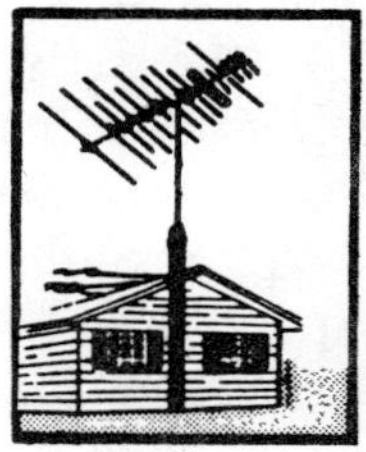

This book is for the home handyman and the neophyte TV technician who wants to know how to select and install television antennas.

Once upon a time, TV antennas were installed almost exclusively by professional TV technicians who knew what kind of an antenna was needed and how to install it. But times are changing. Fewer TV technicians want to climb up a customer's roof and take the chance of being blamed for damaging the roof. Besides, they're busy enough fixing TV sets without having to take on antenna installation work. Most of them are glad to sell antenna system components, but not to install them.

The TV set owner can save money by installing his own antenna system. The cost of TV technician time has gone up. And the cost of TV antennas and associated hardware has gone down. Besides, antennas and accessories are being packaged in "blister bags" and in assemble-it-yourself form.

No technical knowledge is required to understand this book, which explains clearly how to select, where to buy, and how to install TV antennas.

The performance of a TV receiver depends upon its antenna system. What sufficed for a monochrome TV receiver may be very inadequate for color TV.

Devices are available which enable use of single antenna system for two or more TV sets and FM-stereo receivers. Good stereo reception depends upon a good antenna system—all the more so for good quadraphonic (four-channel stereo) reception.

In addition to explaining how to select and install TV antenna system components for use in typical suburban and rural areas, this book contains information for the big-city home and apartment dweller, and even for the weekend sailor who wants to enjoy TV reception on his boat.

The activation of a new TV station in the area might require the installation of a new receiving antenna system to receive the new station clearly. Reception often deteriorates as TV antennas no longer work as well as they did; they are affected by aging and changing conditions. TV antennas do need replacement. And when they are replaced, they should be replaced with new, improved types.

This book makes it clear as to what is needed and how to install it properly.

Leo G. Sands

Contents

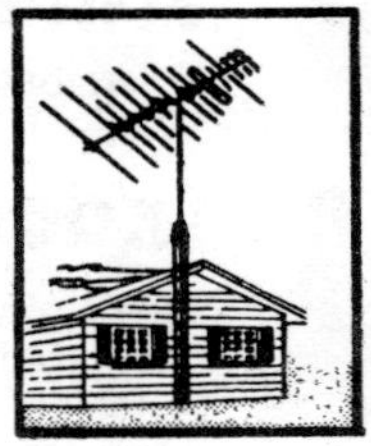

Television Reception

1

The quality of the pictures reproduced on the screen of a television receiver can only be as good as permitted by the antenna that intercepts the television signals. This is all the more true when receiving television programs in color.

PICTURE REPRODUCTION AND TRANSMISSION

Figure 1-1 is a television picture signal flow chart. The television camera converts light variations into an electric signal which is processed and sent through a control console directly to the television station transmitter, or as shown in Fig. 1-1, first to a video tape recorder. The pictures are usually recorded on magnetic tape first so that the program can be

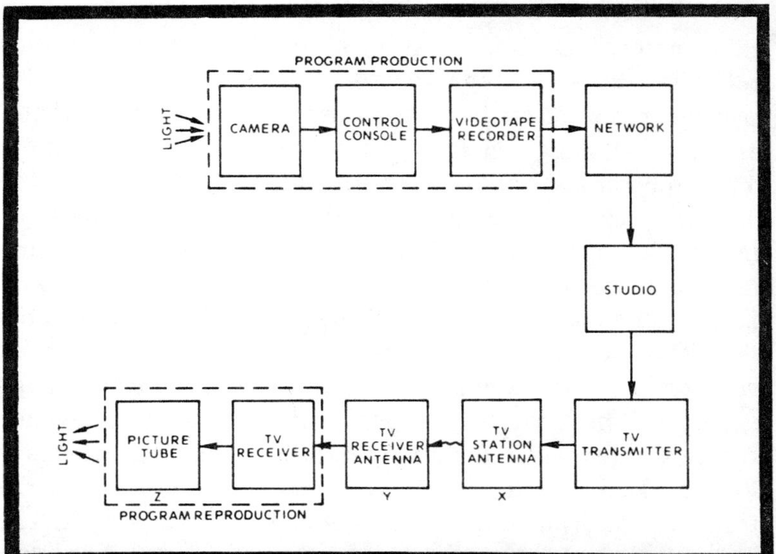

Fig. 1-1. Television picture signal flow chart from camera to picture tube.

Fig. 1-2. A typical color television camera. The operator can see the televised picture in the viewfinder which is a small picture tube. (Courtesy Ampex Corporation.)

edited before being broadcast. Again referring to the diagram, note that the video tape recorder, when playing back the program recorded on the tape, sends its signal to the television network switching center, or directly to the TV transmitter, through a coaxial cable. From the network switching center the program is transmitted through coaxial cables or microwave systems or both to numerous television transmitting stations. The transmitter converts the picture signal into a radio signal which is fed through coaxial cable to the television station transmitting antenna. The radio signal, representing the television picture, is transmitted through space in all horizontal directions for interception by television receiver antennas. In the diagram the transmitting antenna is identified as X and the receiving antenna as Y. It is the transmission from X to Y that is of concern in this book. The signal intercepted by the receiving antenna is fed through a twin-lead or coaxial cable transmission line to the television

Fig. 1-3. NBC television control center in the RCA Building at Radio City in New York. (Courtesy NBC.)

receiver and the picture is reproduced by the picture tube of the receiver. Therefore, it can be seen that the light intercepted by the camera is converted into an electrical signal and is reconverted into light by the picture tube.

In order to faithfully reproduce the picture seen by the camera, it is extremely important that distortion is not introduced anywhere in the chain shown in Fig. 1-1. Furthermore, extraneous signals such as noise should not be added to the picture. And the receiving antenna should receive sufficient energy from the transmitting antenna to enable the television receiver to reproduce the picture with adequate contrast. The television receiving antenna should also receive only one signal from the selected station.

The television program originates with a camera such as the one shown in Fig. 1-2. This is a color television camera whose lens is adjusted by the operator by means of hand grip controls. The operator sees the picture viewed by his or her own camera through a viewer at the back of the camera which is similar to the picture tube of a television receiver. Generally, two or more cameras are used when broadcasting or recording a television program. The outputs of the cameras are fed to a control center such as the one shown in Fig. 1-3. The operator has several television screens which he can watch and, by means of pushbuttons and switches, he can select the one that is going to be recorded for broadcast.

TRANSMITTING AND RECEIVING ANTENNAS

The transmitting antenna of a television broadcast station is usually mounted on top of a tower or tall building. For example, in New York most of the television stations have their antennas on top of the Empire State Building; in Chicago the site is the top of the John Hancock building. In Los Angeles, it's Mt. Wilson, a mountaintop with a clear view of the entire metropolitan area. Some stations, however, have their own individual towers. Figure 1-4 shows a 1000 foot high television antenna tower used by a station in Maine. The higher the television antenna is above the ground and surrounding terrain, the greater its service area.

Fig. 1-4. A 1000-foot tall television station antenna tower. (Courtesy Raytheon Company.)

Fig. 1-5. A typical portable TV receiver having a built-in monopole antenna.

The typical portable television receiver is equipped with a monopole antenna, as illustrated in Fig. 1-5. The monopole antenna can be seen at the rear of the receiver. It is a telescoping rod whose length and position are adjusted for the most clear reception. Such an antenna is satisfactory under some conditions. It is provided by the manufacturer to enable the user to operate the receiver almost anywhere. However, an external antenna system can be connected to the receiver to obtain a much more satisfactory picture reproduction and to enable reception of television signals from greater distances.

Some television receivers are provided with a "rabbit-ear" dipole antenna consisting of two telescoping rods, as illustrated in Fig. 1-6, a rear view of a typical portable television set. Note that in addition to the rabbit ears, there is a small wire loop antenna attached to the back of the receiver for picking up stations that operate in the UHF (ultrahigh-frequency) band. The rabbit ears are used for reception of stations that operate in the VHF (very high-frequency) band. (The stations that operate on channels 2 through 13 are in the VHF band and those that operate on channels 14 through 68 are in the UHF band.) It should be noted in Fig. 1-6 that there are four screw terminals for antenna connections. The rabbit ears are connected internally to the VHF antenna terminals and the UHF loop antenna is connected directly to the UHF antenna terminals. To use an external VHF antenna, the wires from the rabbit ears are disconnected from the VHF antenna terminals and the external antenna is connected through a transmission line as shown in Fig. 1-7.

The simplest type of antenna is shown in this diagram for the sake of simple explanation. The antenna intercepts the signals which are conveyed to the receiver antenna terminals through what is known as a transmission line. In this diagram, only the VHF antenna is shown. When an outdoor UHF antenna is also used, the UHF loop antenna is removed and the UHF antenna is connected to the UHF antenna terminals

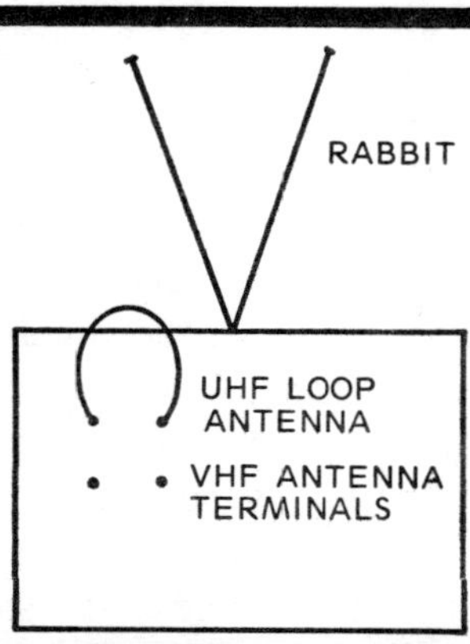

Fig. 1-6. Rear view of a typical TV receiver showing the antenna terminals and build-in antennas. The rabbit ears are connected internally to the VHF antenna input terminals. The UHF loop antenna is connected directly to the UHF input terminals.

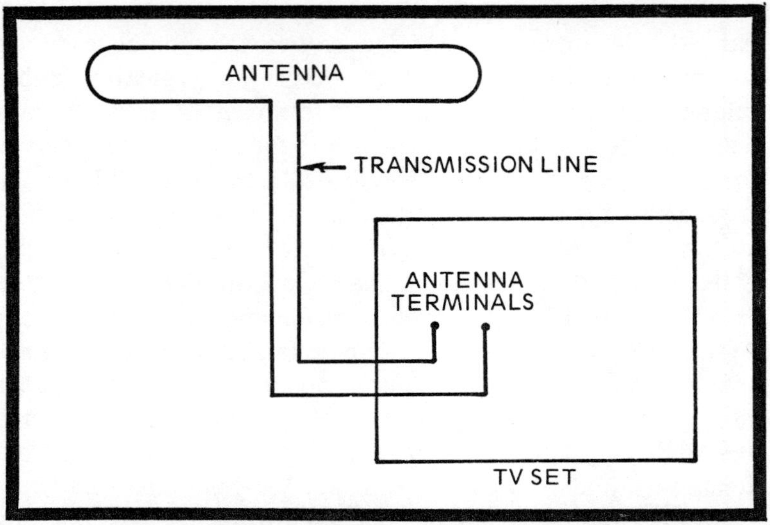

Fig. 1-7. Basic external TV antenna system consisting of the antenna and a transmission line.

through a transmission line in the same manner as the VHF antenna. As will be learned later, a single antenna and transmission line can be used for reception of both VHF band and UHF band television stations.

TELEVISION SIGNAL PROPAGATION

If you throw a rock into a still pool of water, you will see water waves radiating in all directions from the point where the rock hit the water. In the same way, as illustrated in Fig. 1-8, the radio waves radiated by most television broadcast station antennas travel in all directions.

The best reception is obtained when the signal from the transmitting antenna travels through an unobstructed line-of-sight path as shown in Fig. 1-9. As shown here, the receiving antenna can "see" the transmitting antenna. However, even when the receiving antenna can "see" the transmitting antenna, perfect reception will not necessarily result. In Fig. 1-10 it is shown how the signal may travel directly from the transmitting antenna to the receiving antenna and at the same time the signal can travel to some other surface and then be

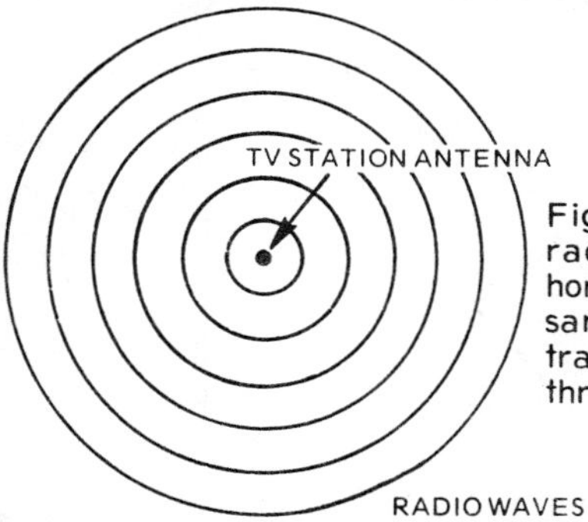

Fig. 1-8. A TV station antenna radiates radio waves in all horizontal directions in much the same manner as water waves travel outward when a rock is thrown into a still pool of water.

reflected back to the receiving antenna. In other words, the television signal travels from X to Z and from X to Y to Z. Since the X-to-Z path is shorter than the X-Y-Z path, the signal arriving via the X-Y-Z path arrives at the receiving antenna later than the signal arriving over the X-Z path. Therefore, when two signals from the same station are received at the receiving antenna via two paths of differing lengths, "ghosty" pictures with multiple images will result.

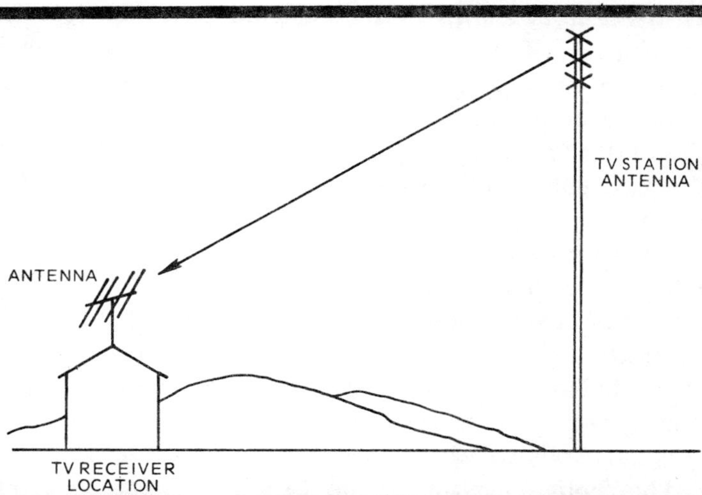

Fig. 1-9. Exaggerated example of line-of-sight transmission. The receiving antenna can literally see the transmitting antenna.

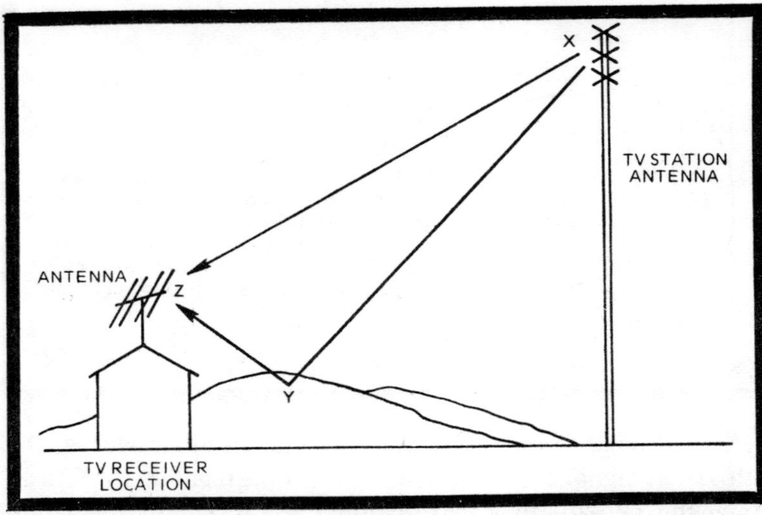

Fig. 1-10. Example of multipath reception. The TV signal travels directly from the transmitting antenna X to the receiving antenna Z. It can also travel from X to the ground at Y from where it is reflected to Z.

In many cases, particularly in large cities, the receiving antenna cannot "see" the transmitting antenna. Reception in such cases is possible via a reflected path as shown in Fig. 1-11. Here it is shown that the signal from the television station, point X, travels toward the receiving antenna but encounters an obstacle. The signal from X also travels to another large obstacle to the side, point Y, from where it is reflected to the receiving antenna, point Z. The receiving antenna is pointed at Y so it can pick up the reflected signal from X.

As is well known, television stations can be received at distances beyond the horizon. As shown in Fig. 1-12, an example of beyond-the-horizon reception, the signal from the transmitting station X travels to the horizon, point Y, where it is refracted and reaches the receiving antenna at point Z. The signal at point Z, however, is not anywhere near as strong as the signal at point Y where the signal strikes the horizon.

The "optical horizon" is the distance one can see from a given point. For example, when standing on the observatory floor of the Empire State Building, it is possible in clear

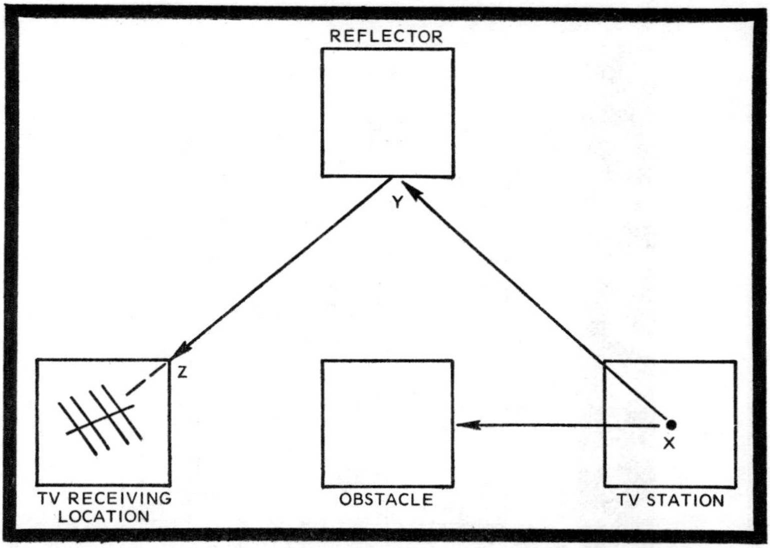

Fig. 1-11. Example of reflected path reception. Because of the obstacle in the direct path between TV station X and receiving antenna Z, the signal travels to reflecting surface Y which reflects the signal to Z.

weather to see for a great distance. The "radio horizon" is farther away than the optical horizon. As Fig. 1-13 shows, the line-of-sight distance depends upon the height of the receiving and transmitting antennas. If the transmitting antenna is 1000 feet above the ground and the radio horizon is 50 miles away when the earth is smooth, good TV reception could be expected with the receiving antenna at ground level provided there are no obstacles in the path. However, this condition seldom exists. If the transmitting antenna height is 1000 feet and the receiving antenna height is 20 feet, good television reception

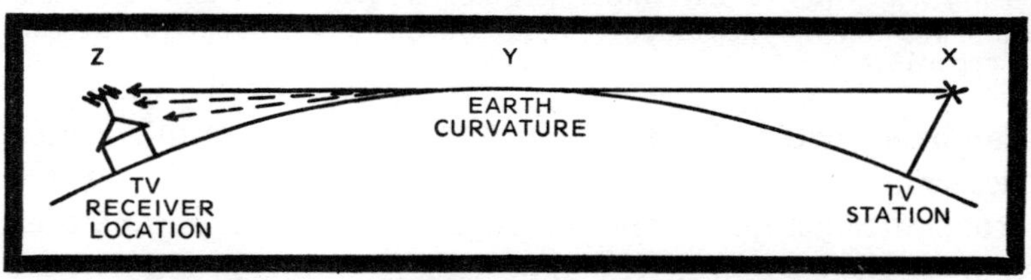

Fig. 1-12. Reflection beyond the horizon. Because of the earth's curvature, the distant receiver at Z intercepts the signal from TV station X because of refraction of the signal at the horizon, point Y.

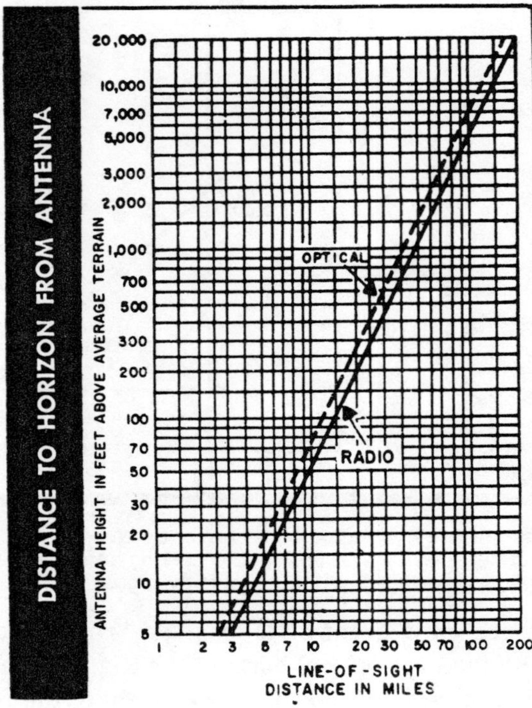

To determine line-of-sight distance between two antennas of different height: Use chart to determine line-of-sight distance for each and then add the two distances.

Fig. 1-13. Radio horizon graph. The line-of-sight distance between the transmitting antenna and the receiving antenna can be determined by using the graph to determine the antenna-to-horizon distance of each antenna and by then adding these two distances. (Courtesy Jerrold Corporation.)

can be expected at a distance of 57 miles. To use this graph, determine the line-of-sight distance to the horizon for the transmitting antenna and for the receiving antenna and then add the two distances. It can be seen from this graph that the higher the receiving antenna, the greater the distance from which television stations can be received.

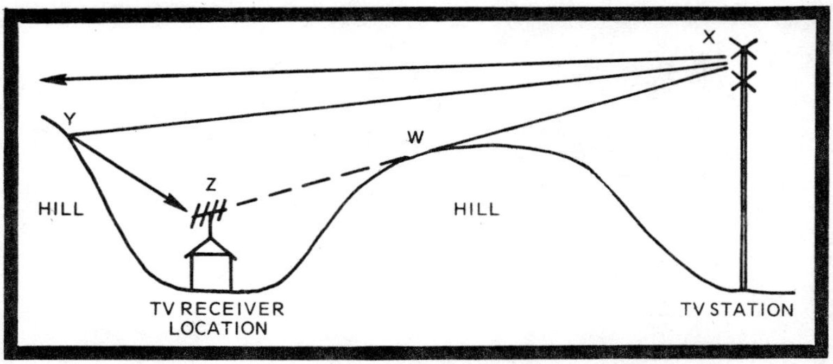

Fig. 1-14. Reception in a valley may be poor or impossible, particularly if the distance from the transmitter to the receiver is great. It is possible that the refracted energy from point W will reach the receiving antenna at Z. If another hill beyond Z exists and is in line-of-sight of X, the signal might be reflected from Y to Z.

RECEPTION PROBLEMS

When hills intervene between the transmitting antenna and the receiving antenna, as shown in Fig. 1-14, it may not be possible to receive signals from the transmitting station. In this diagram it is shown that the signal from X strikes the top of the hill, point W, and might be refracted down to the receiving station location, point Z. It depends upon the distance from X to W and the height of W with respect to Z if any reception is to be obtained. On the other hand, if the signal

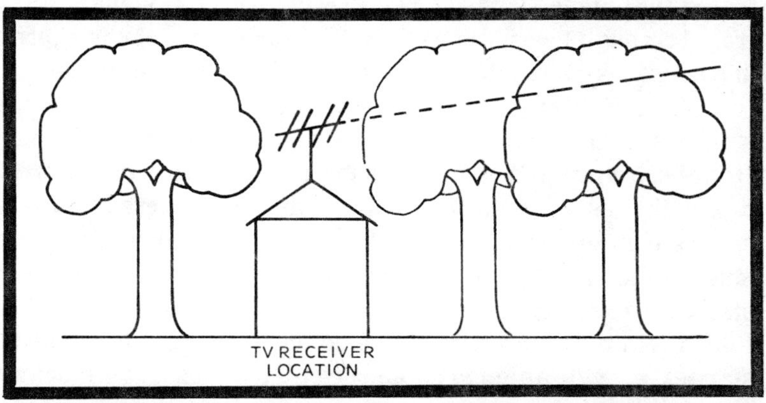

Fig. 1-15. Trees absorb TV signals. If the antenna does not extend above the trees, significant absorption losses can result.

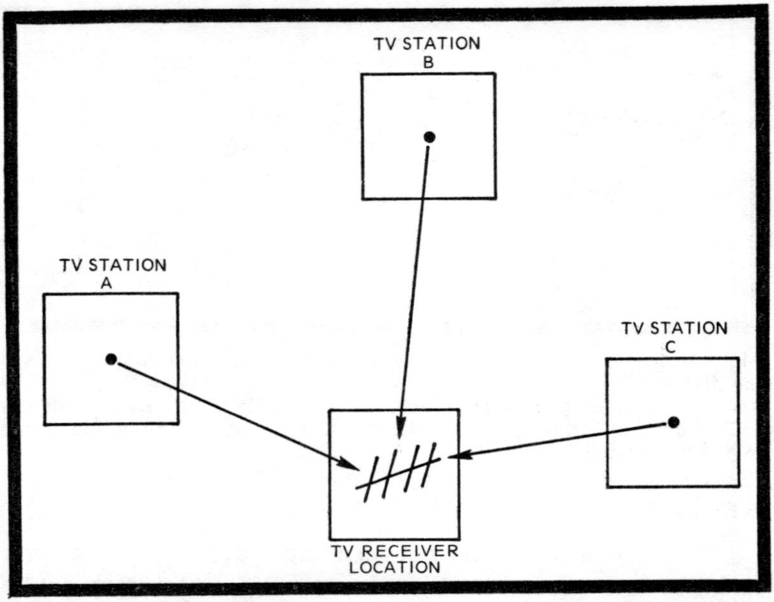

Fig. 1-16. Reception of TV stations in different directions may require the use of more than one antenna or a rotator.

from X strikes a hill on the other side of the receiving location, point Y, and reflects a signal back to point Z, reception will be possible.

The waves transmitted by a television antenna are affected by foliage. If, as shown in Fig. 1-15, the signals must travel through trees, the trees will absorb some of the energy and cause less of it to be received by the receiving antenna.

When the stations to be received are all in one direction, a single antenna can be installed and locked into position. However, when the stations to be received are in different directions, as shown in Fig. 1-16, it may be necessary to use separate antennas for receiving from different directions or a single antenna mounted on a rotator. When switching from station A to station B for example, the antenna is rotated so that it will point towards B and so on. However, in some instances, a single antenna can be oriented so that it will pick up direct signals from some stations and reflected signals from other stations.

RADIO PROPAGATION LOSS

Television stations broadcast on frequencies in the VHF and UHF portions of the radio spectrum as illustrated in Fig. 1-17. Television channels 2 through 6 are located within the 54-88 MHz (megahertz) frequency range. Channels 7 through 13 are within the 174-216 MHz range. And the UHF television channels 14 through 68 are within the 470-800 MHz range. In some localities, channels 14 through 20 (between 470 and 512 MHz) are no longer available for television broadcasting and are instead used for two-way radio communications.

The term "megahertz" means frequency in millions of cycles per second. The waves transmitted by the television station are extremely short, and possess many of the properties of light. The higher the frequency, the shorter the wavelength, and the more the signal is affected by distance. As shown in Fig. 1-18, the signal from a station on channel 2 is ten times stronger than the signal from a station on channel 68 when both stations have antennas of equal height and have the same effective radiated power. It is for this reason that the UHF stations cannot be received at as great a distance as VHF stations.

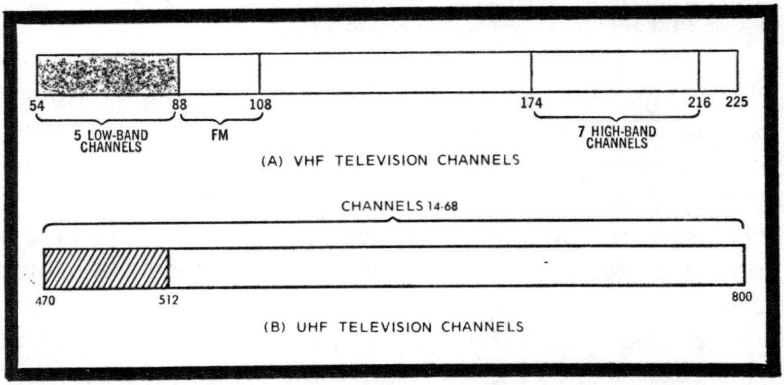

Fig. 1-17. The television broadcast channels are located within the VHF and UHF portions of the radio spectrum. Channels 2 through 6 are in the 54-88 MHz low VHF band; channels 7 through 13 are in the 174-216 MHz high VHF band; and channels 14 through 68 are in the 470-800 MHz UHF band. Channels 14 through 20 within the 470-512 MHz portion of the UHF band are not available for TV broadcasting in some areas. Channels 68 through 83 above 800 MHz have been reallocated for other uses.

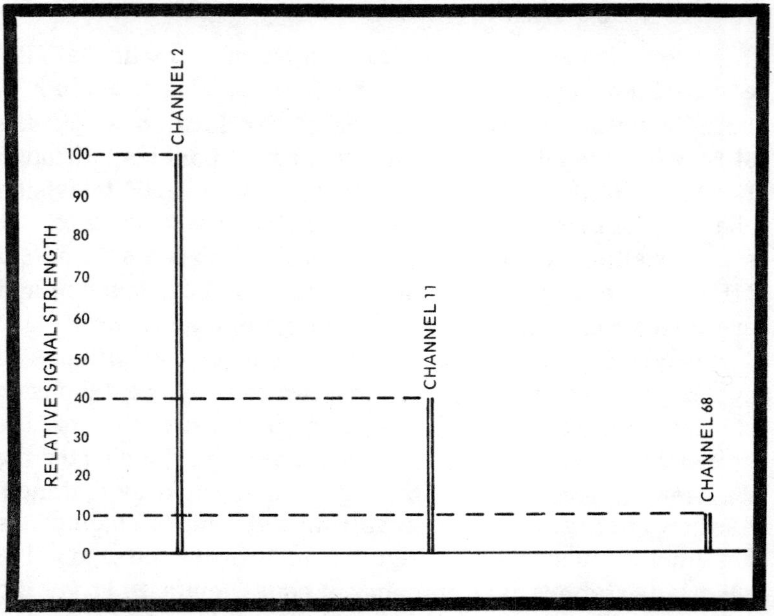

Fig. 1-18. The transmission loss through space rises with frequency as indicated above. If stations operating on channels 2 and 68 radiate the same amount of power, have antennas of the same height and are at equal distances from the receiving antenna, the channel 2 signal will be 10 times stronger than the channel 68 signal.

As the table of Fig. 1-19 indicates, the signal from a television station actually consists of three signals, the video signal (picture), the color signal, and sound signal. Each television channel is 6 MHz wide, as indicated in the table. It is necessary for the antenna to capture all three signals of a color television program in order to reproduce the picture clearly.

Actually, a television station transmitting antenna radiates two radio signals simultaneously. One of these radio signals contains the picture and color information. The other radio signal contains the sound information.

In the following chapters you will learn about characteristics of antennas and transmission lines and how to design and install your own antenna system.

CH	Freq. Range MHz	Carriers in MHz Video	Color	Sound	Wave lengths in inches Air *
LOW VHF		–	–	–	–
2	54-60	55.25	58.83	59.75	207.0
3	60-66	61.25	64.83	65.75	187.0
4	66-72	67.25	70.83	71.75	171.0
5	76-82	77.25	80.83	81.75	149.0
6	82-88	83.25	86.83	87.75	139.0
FM	88-108	–	–	–	–
HIGH VHF		–	–	–	–
7	174-180	175.25	178.83	179.75	66.7
8	180-186	181.25	184.83	185.75	64.5
9	186-192	187.25	190.83	191.75	62.5
10	192-198	193.25	196.83	197.75	60.5
11	198-204	199.25	202.83	203.75	58.8
12	204-210	205.25	208.83	209.75	57.1
13	210-216	211.25	214.83	215.75	55.5
UHF		–	–	–	–
14	470-476	471.25	474.83	475.75	25.0
15	476-482	477.25	480.83	481.75	21.7
16	482-488	483.25	486.83	487.75	–
17	488-494	489.25	492.83	493.75	–
18	494-500	495.25	498.83	499.75	–
19	500-506	501.25	504.83	505.75	–
20	506-512	507.25	510.83	511.75	23.2
21	512-518	513.25	516.83	517.75	–
22	518-524	519.25	522.83	523.75	–
23	524-530	525.25	528.83	529.75	–
24	530-536	531.25	534.83	535.75	–
25	536-542	537.25	540.83	541.75	21.9
26	542-548	543.25	546.83	547.75	–
27	548-554	549.25	552.83	553.75	–
28	554-560	555.25	558.83	559.75	–
29	560-566	561.25	564.83	565.75	–
30	566-572	567.25	570.83	571.75	20.8
31	572-578	573.25	576.83	577.75	–
32	578-584	579.25	582.83	583.75	–

CH	Freq. Range MHz	Carriers in MHz Video	Color	Sound	Wave * lengths in inches Air
33	584-590	585.25	588.83	589.75	–
34	590-596	591.25	594.83	595.75	–
35	596-602	597.25	600.83	601.75	19.7
36	602-608	603.25	606.83	607.75	–
37	608-614	609.25	612.83	613.75	–
38	614-620	615.25	618.83	619.75	–
39	620-626	621.25	624.83	625.75	–
40	626-632	627.25	630.83	631.75	18.8
41	632-638	633.25	636.83	637.75	–
42	638-644	639.25	642.83	643.75	–
43	644-650	645.25	648.83	649.75	–
44	650-656	651.25	654.83	655.75	–
45	656-662	657.25	660.83	661.75	17.9
46	662-668	663.25	666.83	667.75	–
47	668-674	669.25	672.83	673.75	–
48	674-680	675.25	678.83	679.75	–
49	680-686	681.25	684.83	685.75	–
50	686-692	687.25	690.83	691.75	17.1
51	692-698	693.25	696.83	697.75	–
52	698-704	699.25	702.83	703.75	–
53	704-710	705.25	708.83	709.75	–
54	710-716	711.25	714.83	715.75	–
55	716-722	717.25	720.83	721.75	16.4
56	722-728	723.25	726.83	727.75	–
57	728-734	729.25	732.83	733.75	–
58	734-740	735.25	738.83	739.75	–
59	740-746	741.25	744.83	745.75	–
60	746-752	747.25	750.83	751.75	15.8
61	752-758	753.25	756.83	757.75	–
62	758-764	759.25	762.83	763.75	–
63	764-770	765.25	768.83	769.75	–
64	770-776	771.25	774.83	775.75	–
65	776-782	777.25	780.83	781.75	15.2
66	782-788	783.25	786.83	787.75	–
67	788-794	789.25	792.83	793.75	–
68	794-800	795.25	798.83	799.75	–

* All measurements at mid-band.

Fig. 1-19. Each TV "signal" is actually a broad spectrum (6 MHz) comprised of several individual signals, particularly these three basic ones: video, color information, and audio.

Television Receiving Antennas

2

The function of the television receiving antenna is to intercept the radio waves radiated by the television transmitting antenna. The electromagnetic energy in the air that is radiated by the transmitting antenna is intercepted by the antenna and fed to the TV receiver through the transmission line.

BASIC ANTENNAS

The simplest television antenna is called the "dipole," of which an example is shown in Fig. 2-1. Here, the antenna consists of two lengths of copper tubing of equal length with one end of each attached to an insulating block. This kind of antenna is mounted so that its rods are parallel to the earth and face the direction from which the television signal is coming. This kind of antenna is seldom used because it is most sensitive only at a particular channel. The channel to which it is most receptive depends upon the length of the rods, which corresponds to portions of the physical length of the radiated wave. For example, to receive channel 2, each rod would be approximately 4.5 feet long (¼ wavelength). The overall length from X to Z therefore would be approximately nine feet. Another shortcoming of this type of antenna is the fact that it is as receptive to signals from in front of it as from behind it. It is least sensitive to stations in the direction toward which the ends of the antenna are pointed.

A folded dipole antenna is shown in Fig. 2-2. It is similar to the dipole except that it consists of a pair of "hairpin" loops. As shown in the diagram, it may consist of a bent length of copper tubing attached to an insulating block at point Y. The length of this antenna is the same as that of a dipole, as are its directional characteristics. It receives equally as well from its

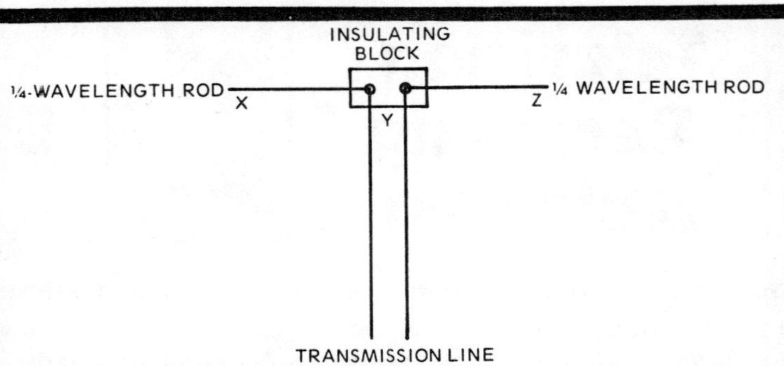

Fig. 2-1. Basic half-wave dipole antenna consisting of two quarter-wave elements.

broadside directions, front and back. An antenna that is receptive to signals from all directions is illustrated in Fig. 2-3. This is a dual folded dipole (or turnstile). Because of its omnidirectional characteristics, it is seldom used for television reception because it would pick up not only the desired signal but many reflected signals. This type of antenna, however, is widely used for FM broadcast band radio reception.

PRACTICAL ANTENNAS

The folded-dipole antenna, previously shown in Fig. 2-2, can be improved by adding another rod behind the folded dipole as shown in Fig. 2-4. The rod behind the dipole acts as a

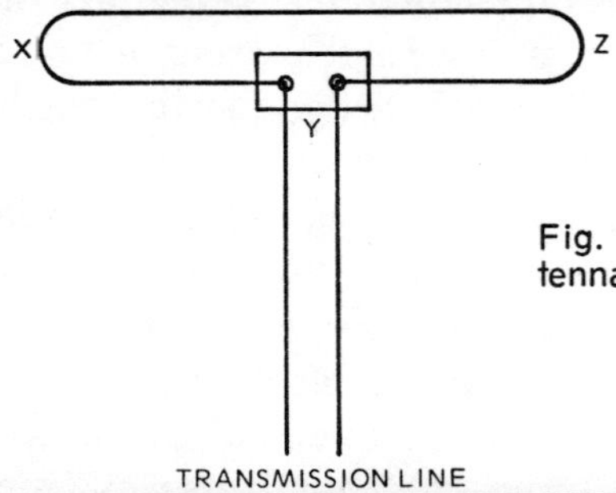

Fig. 2-2. Basic folded dipole antenna.

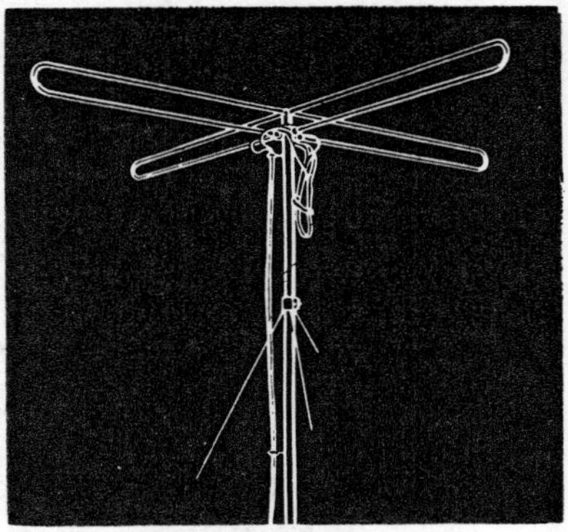

Fig. 2-3. Dual folded dipole antenna for omnidirectional reception.

signal "reflector." The antenna picks up signals and the reflector also picks up signals which are reflected back toward the antenna and thus increases the sensitivity of the antenna. The reflector also minimizes reception from the rear.

Although it was stated that the dipole and folded-dipole antennas provide optimum reception on one channel only, they are capable of picking up signals on several channels. Generally, the folded dipole is cut to such a length that it will pick up the low-band VHF television channels (2-6) or the high-band VHF channels (7-13). Figure 2-5 shows a dual-band antenna consisting of a high-band VHF folded dipole in front and a low-band dipole behind. Behind both of the dipoles is a reflector. Again the reflector improves the antenna sensitivity and minimizes reception from the rear.

The sensitivity and the backside rejection capabilities of such an antenna can be improved by stacking two such antennas, one above the other, as shown in Fig. 2-6.

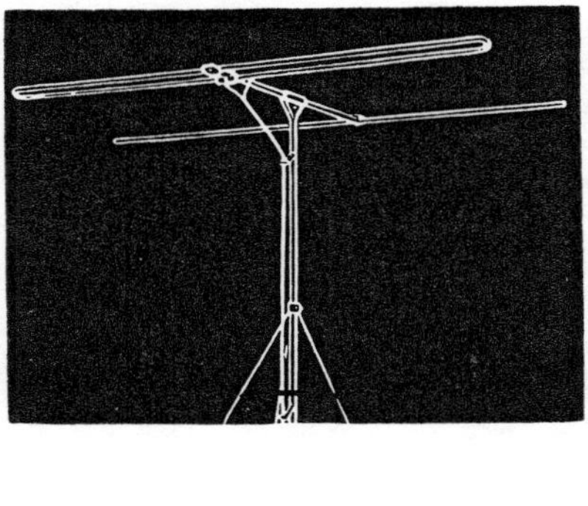

Fig. 2-4. Folded dipole with reflector.

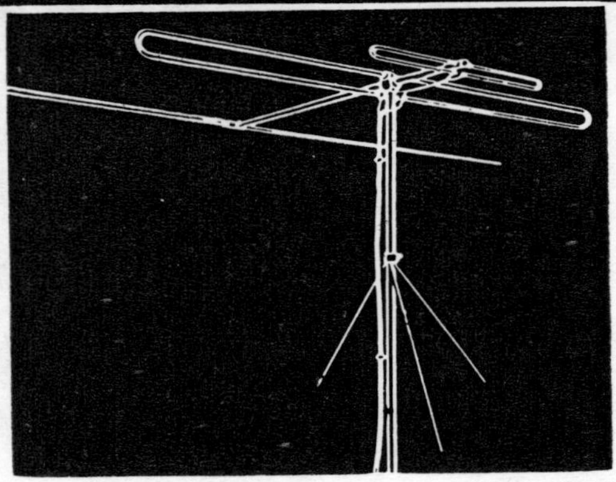

Fig. 2-5. Low—high band antenna consisting of two folded dipoles and a reflector.

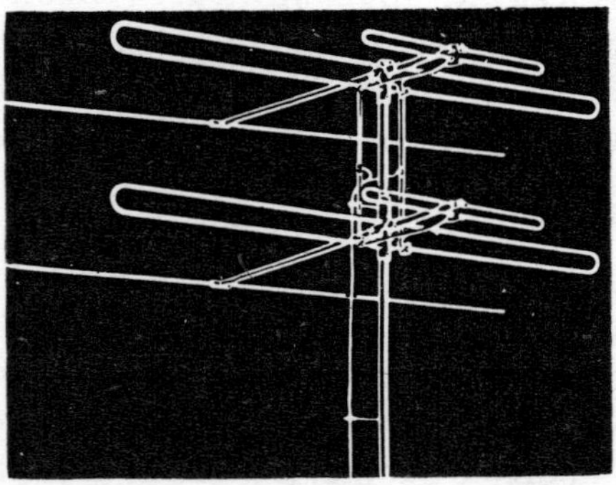

Fig. 2-6. Stacked array consisting of two antenna of the same type as previously shown in Fig. 2-5.

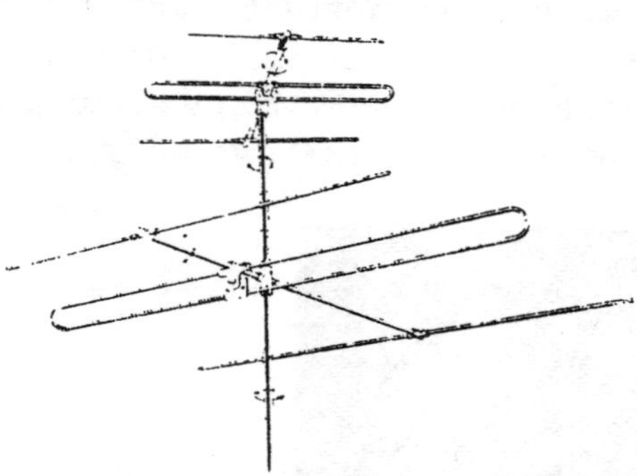

Fig. 2-7. Dual-band antenna system consisting of two independently adjustable yagi arrays.

received are in the same direction. Figure 2-7 shows an antenna consisting of two independent antennas, a high-band antenna on top and a low-band antenna below. Each of these antennas consists of a dipole and a reflector plus a third elment in front known as a "director." This type of antenna is known as a yagi array. The antenna shown here can be adjusted to provide optimum reception on low-band channels from one direction and on high-band channels from another direction.

UHF Antennas

As was noted previously in Fig. 1-19, the frequencies of the UHF television channels are much higher than those of the VHF television channels. This means that the lengths of the radio waves are much shorter and that much smaller antennas can be used. The antennas previously shown in this chapter are not suitable for UHF band reception unless they are correctly dimensioned.

One of the very practical UHF antennas is illustrated in Fig. 2-8. This is commonly known as the bow-tie antenna. The dipole element consists of two triangular pieces of metal placed in front of a grid-like reflector. It is highly directional and the rejection of signals from the rear is greatly improved. Much greater directivity and sensitivity can be achieved by

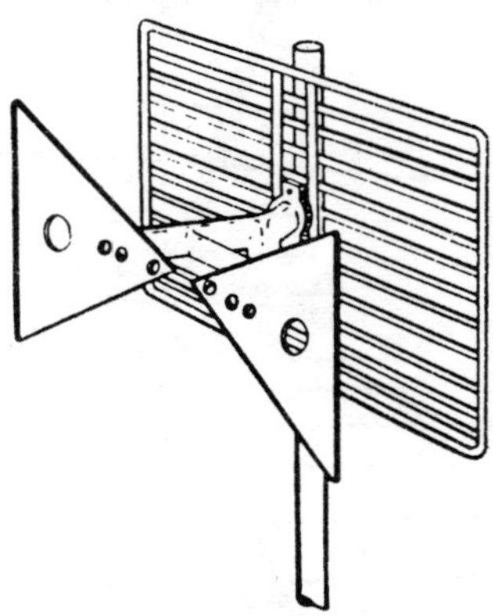

Fig. 2-8. Bow-tie antenna for the UHF band. (Courtesy GC Electronics.)

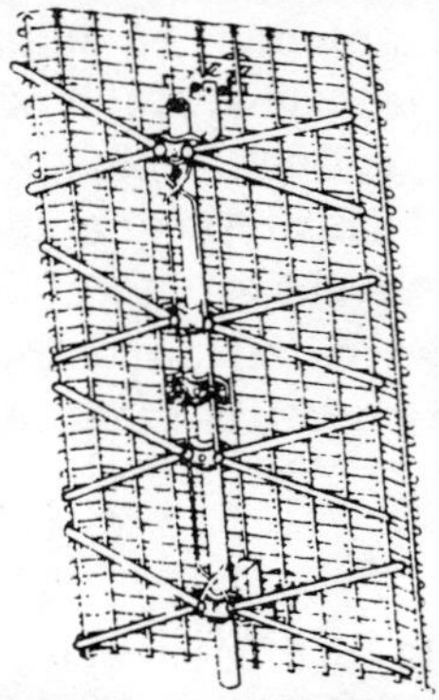

Fig. 2-9. Four-bay quadri-bow antenna for the UHF band. (Courtesy GC Electronics.)

stacking bow-tie antennas as shown in Fig. 2-9. In this type of antenna there are four bow ties consisting of metal rods placed in front of a grid-like reflector.

Another practical UHF antenna is shown in Fig. 2-10. This is a corner reflector type of antenna which consists of a bow tie backed by two reflectors. Each of the reflectors is at an angle with respect to the bow tie.

When maximum directivity is required, the parabolic antenna illustrated in Fig. 2-11 is commonly used. It works on

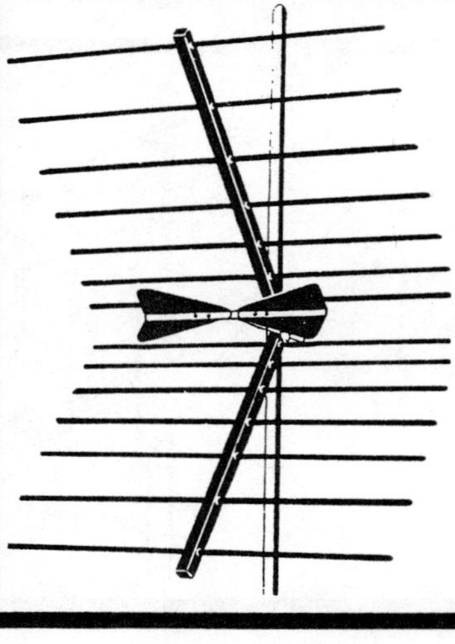

Fig. 2-10. Corner reflector antenna employing a bow-tie as the dipole element. (Courtesy Channel Master.)

the theory of an automobile headlight or a spotlight. The folded dipole is held in front of the parabolic reflector. The signals are first intercepted by the reflector and then reflected back to the antenna.

Another example of a UHF antenna is shown in Fig. 2-12. This antenna utilizes a corner reflector with rods of varying lengths with four folded dipoles within the reflector assembly. A more exotic UHF antenna is shown in Fig. 2-13. As can be seen, it consists of a large number of elements. This type of antenna is highly directive and sensitive.

VHF Antennas

It is seldom necessary to use separate low-band and high-band VHF antennas when all the stations to be received are in the same direction. There are numerous antennas that are designed to cover both bands as well as the FM radio broad-

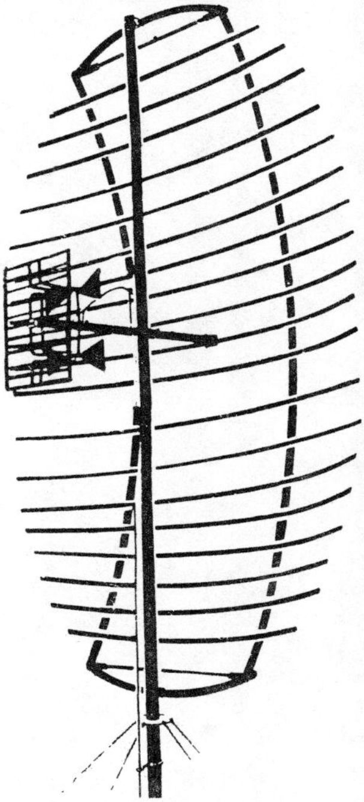

Fig. 2-11. Parabolic antenna for the UHF band. (Courtesy Channel Master.)

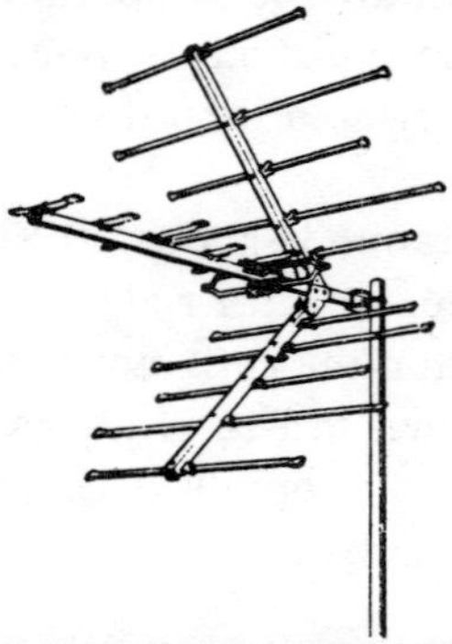

Fig. 2-12. Corner reflector antenna with multiple dipoles. (Courtesy Winegard Antenna Systems.)

cast band which lies between the low and high VHF television bands. Figure 2-14 shows a typical VHF antenna designed for reception of stations up to 150 miles away. It consists of 19 elements and has high directivity and sensitivity.

Another VHF antenna is shown in Fig. 2-15. This antenna employs the "log periodic" principle. The lengths of the

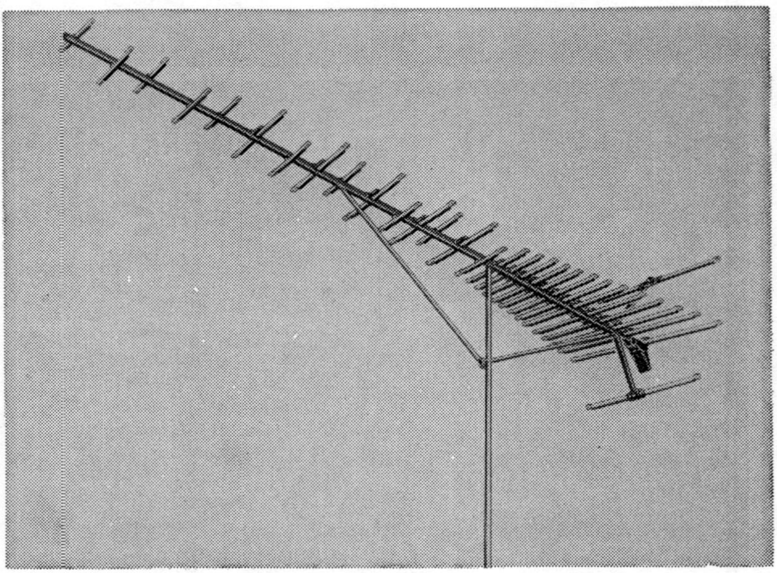

Fig. 2-13. UHF band antenna for long-range reception. (Courtesy Winegard Antenna Systems.)

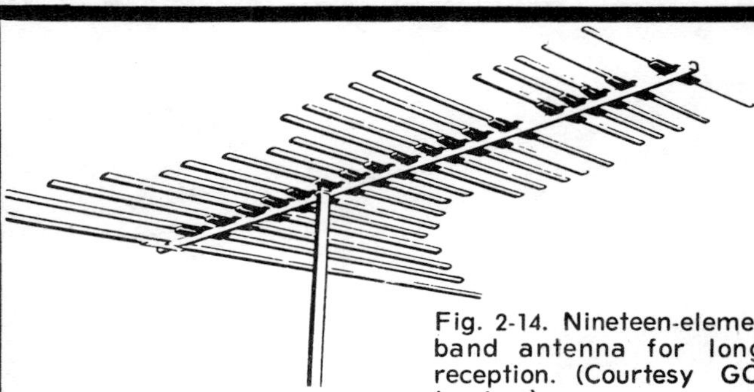

Fig. 2-14. Nineteen-element VHF band antenna for long-range reception. (Courtesy GC Electronics.)

elements vary so that the antenna is frequency-independent throughout its intended range. This particular antenna has 19 active elements and is 205 inches long. It is claimed to be capable of picking up signals from distances up to 225 miles.

VHF—UHF Antennas

There are many antennas on the market that are designed for reception of both VHF and UHF television channels. Examples of three antennas of this type are shown in Fig. 2-16. The 6-element antenna shown at the top is designed for metropolitan area reception of up to 50 miles in the VHF band and up to 35 miles in the UHF band. The 7-element antenna shown in the middle has a rated maximum range of 50 miles in the VHF band and 40 miles in the UHF band. The 11-element

Fig. 2-15. Log periodic antenna for the VHF band made by JFD. (Courtesy Lafayette Radio Electronics Corp.)

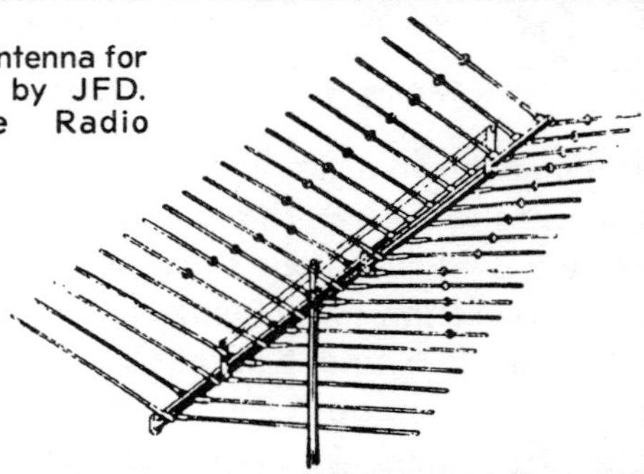

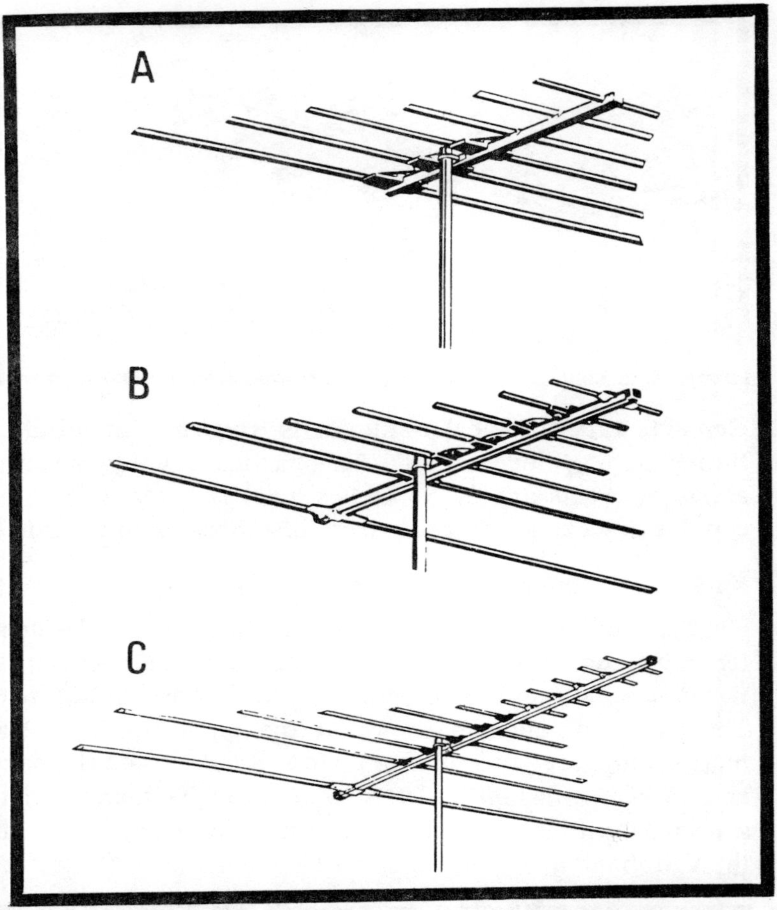

Fig. 2-16. Examples of all-channel antennas. (A) 6-element for local reception. (B) 7-element for medium-range reception. (C) 11-element for use in fringe areas. (Courtesy GC Electronics.)

antenna shown at the bottom has a rated maximum range of 100 miles in the VHF band and 55 miles in the UHF band. For extremely long-range reception, even larger antennas are used. Two examples of long-range antennas are shown in Fig. 2-17. The 19-element antenna shown at the top has a rated maximum range of 125 miles in the VHF band and 75 miles in the UHF band. The 24-element antenna shown at the bottom has a rated maximum range of 175 miles in the VHF band and

90 miles in the UHF band. Both of these antennas are also suitable for FM radio reception.

Other examples of exotic all-channel antennas are shown in Figs. 2-18 and 2-19. As can be noted, they employ a variety of elements of different lengths. The lengths and spacings of the elements are critical in order to achieve the desired directivity and adequate sensitivity. There are so many hundreds of different models of all-channel TV antennas that space does not permit illustrating and discussing more than just a few examples.

SINGLE-CHANNEL ANTENNAS

A single-channel antenna may be a single dipole or folded dipole, as previously shown in Fig. 2-1 and 2-2. However, neither of these antennas has the directivity and sensitivity of specially designed single-channel antennas such as the one

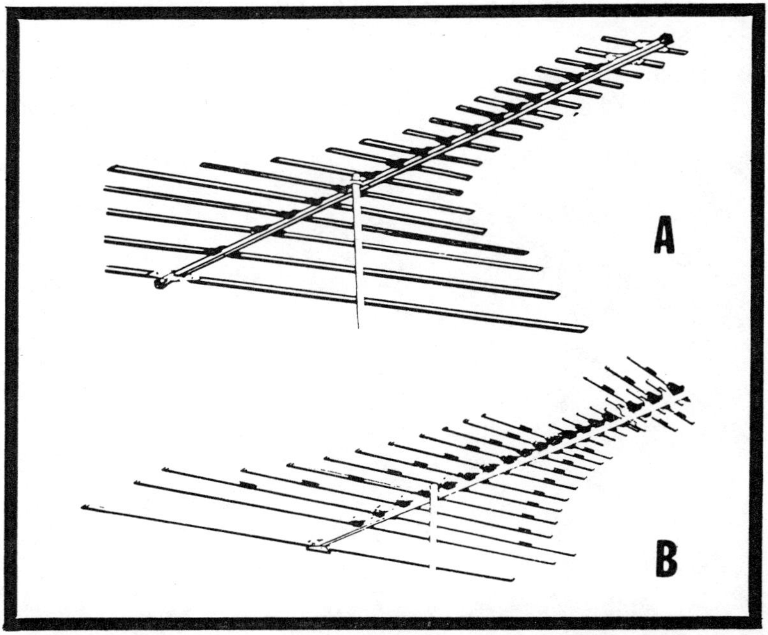

Fig. 2-17. Examples of long-range all-channel antennas. (A) 19-element for far fringe reception. (B) 24-element for deep-fringe reception. (Courtesy GC Electronics.)

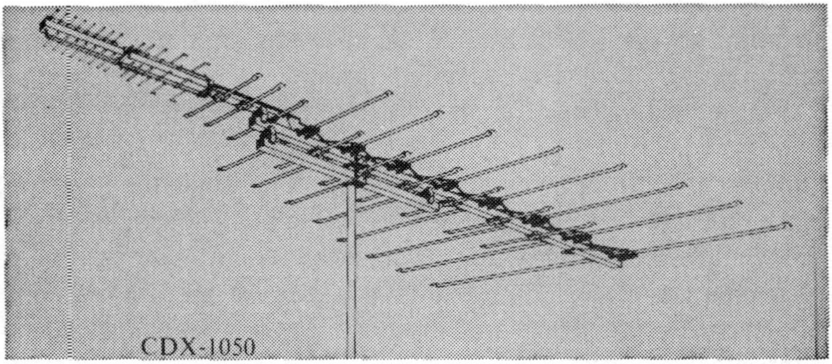

Fig. 2-18. All-channel antenna for VHF, UHF and FM reception. (Courtesy Allied Radio Shack.)

shown in Fig. 2-20. This one is an 11-element yagi which has extremely good directional characteristics and sensitivity. Its elements are cut to the required length, at the factory, for optimum reception of a specific TV channel.

Such an antenna would be used where reception of only one television station is required, particularly from a significant distance. Such an antenna would also be used where all other stations to be received are in a different

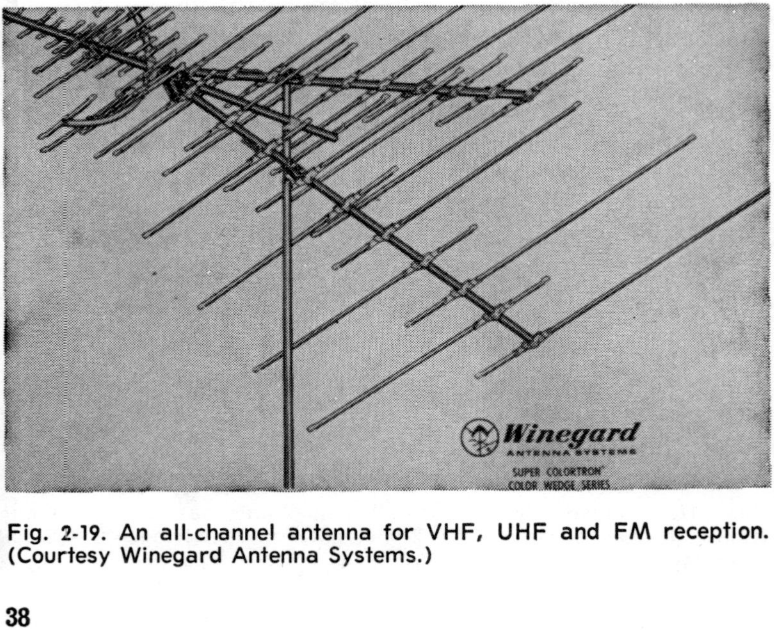

Fig. 2-19. An all-channel antenna for VHF, UHF and FM reception. (Courtesy Winegard Antenna Systems.)

direction, in which case an all-channel and a single-channel antenna would be used.

For reception of an extremely distant station or a very poor receiving location, two single-channel antennas can be stacked, one above the other (Fig. 2-21) or side by side (Fig. 2-22). Even greater sensitivity can be achieved by stacking four antennas as shown in Fig. 2-23.

Single-channel antennas can be used in combination in order to receive extremely distant stations and nearby stations in differing directions. Figure 2-24 shows three antennas attached to the same mast. The one at the top, in this example, is a high-band yagi designed for optimum reception on channels 9 and 11. In the middle is a yagi designed for best reception on channel 5. The antenna at the bottom is an all-channel type. Each of the antennas can be positioned individually.

ANTENNA SENSITIVITY

So far, reference has been made to antenna sensitivity in very broad terms. While the consumer is seldom concerned

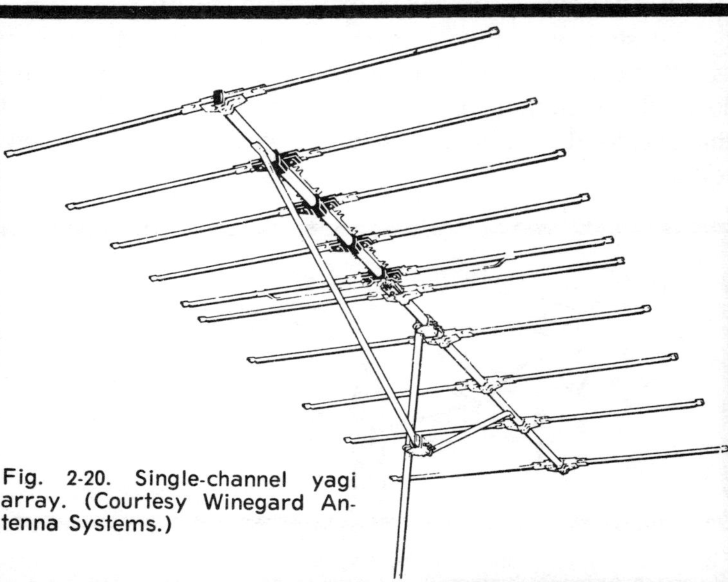

Fig. 2-20. Single-channel yagi array. (Courtesy Winegard Antenna Systems.)

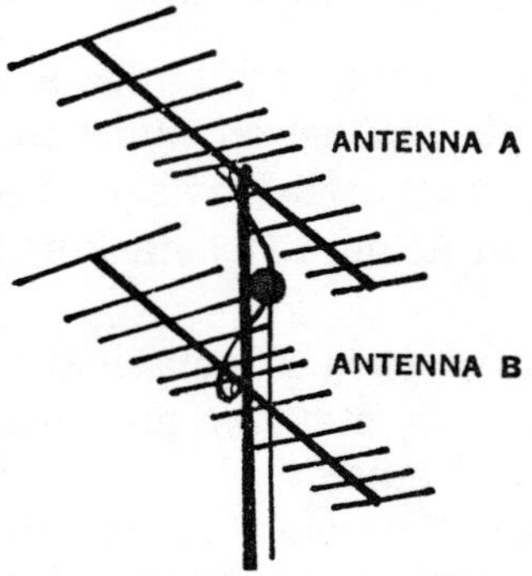

Fig. 2-21. Single-channel antennas stacked one above the other. (Courtesy Winegard Antenna Systems.)

with quantitative terms in regard to the efficiency of an antenna, the engineer is very much concerned. Instead of "sensitivity," the engineer refers to antenna effectiveness in terms of "antenna gain" and "front-to-back ratio" expressed in decibels (dB).

The gain of a dipole or folded dipole is a reference point for comparing other antennas. Thus, the dipole's gain is said to be "unity," or "one." This can be expressed as 0 dB.

TV signals at the receiver can be measured. They are usually very weak, in terms of actual measurement, but they do have an actual voltage level. Stronger signals will be characterized by a higher voltage; weaker ones by a lower voltage.

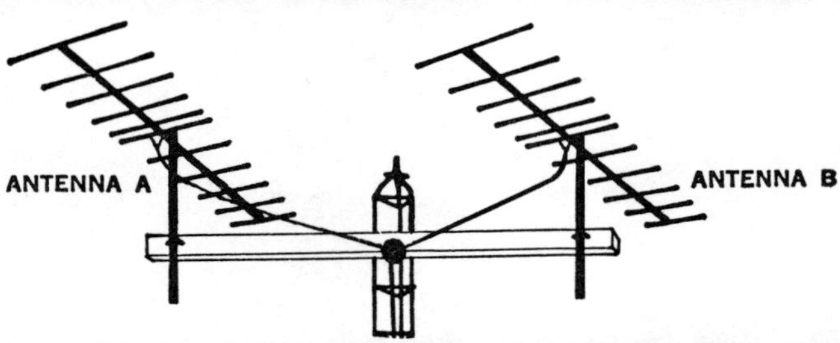

Fig. 2-22. Single-channel antennas stacked side by side. (Courtesy Winegard Antenna Systems.)

Voltage increases or decreases can be expressed in decibels, with 6 decibels (dB) representing the point at which any voltage actually doubles. Two volts is 6 dB above 1 volt. One-half volt is 6 dB below 1 volt.

Antenna signals, of course, are too weak to be expressed in volts. Instead, they have a level equal to perhaps a few hundred millionths of a volt. So a microvolt, or millionth of a volt, is used as the unit for measuring antenna signals.

If antenna gain is 6 dB, it means that the signal voltage from that antenna will be twice that of a reference dipole.

The gain of the 11-element yagi shown in Fig. 2-20 can be from 9.3 to 13 dB, depending upon the channel for which its elements have been cut. If its gain at its selected channel is 12 dB, for example, it will deliver four times more signal voltage to the TV receiver than a simple dipole. If the dipole antenna will cause the level of the signal at the receiver antenna terminals to be 100 microvolts, the 12 dB gain antenna

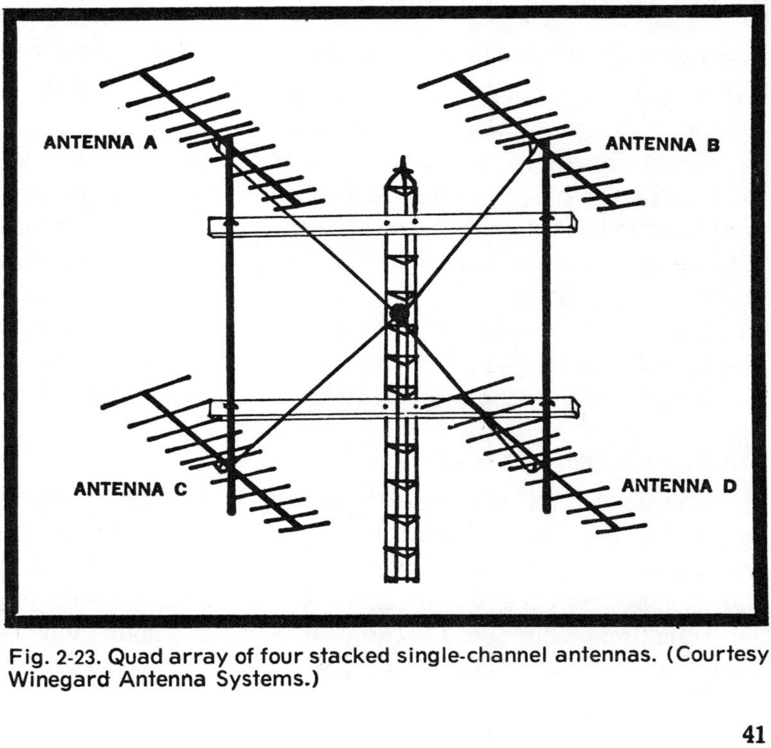

Fig. 2-23. Quad array of four stacked single-channel antennas. (Courtesy Winegard Antenna Systems.)

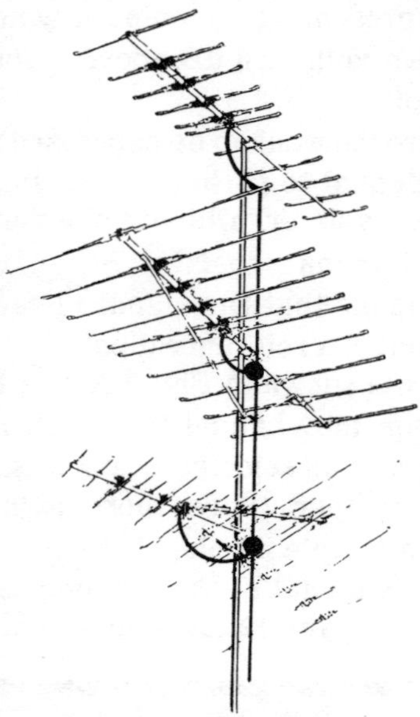

Fig. 2-24. Triple antenna system. The antenna at the top is for the high band VHF channels. The one in the middle is a single-channel antenna and the one at the bottom is an all-channel antenna. (Courtesy Winegard Antenna Systems.)

will cause the signal level to be 400 microvolts, which is adequate for good color TV reception.

Front-to-back ratio is also expressed in dB. The front-to-back ratio of a dipole is 0 dB. But the front-to-back ratio of a sophisticated antenna can be as high as 40 dB. This means that if there is a TV station in the forward direction and another in the directly opposite direction, both are the same distance away and both transmit as much power on the same channel, the signal from the station in the forward direction will be 100 times stronger than the signal from the reverse direction.

ANTENNA DIRECTIVITY

The directional characteristics of a TV antenna are expressed in degrees of "beamwidth." The beamwidth of the yagi shown in Fig. 2-20 ranges from 61 degrees when it is dimensioned for reception on channel 2 to 36 degrees when cut for reception on channel 13. The beamwidth of single-channel UHF antennas is much narrower.

The beamwidth of all-channel antennas varies with frequency and the design of the antenna. Generally, the higher the frequency (channel number), the narrower the bandwidth. This means that the antenna usually has to be aimed more accurately at the source of the highest channel to be received than at lower channel stations.

Beamwidth is particularly important at a location where TV signals can arrive via two or more paths. By making the beamwidth narrow, the antenna will be able to pick up maximum energy from the desired direction and provide maximum rejection of signals coming from undesired directions.

SIGNAL LEVELS

The level of the TV signal at the receiver antenna terminals must be higher than 100 microvolts for satisfactory reception. It is often higher and depends upon the radiated power of the TV station, transmission losses between the transmitting and receiving antennas, the antenna gain and the transmission line losses. The signal level can also be expressed in terms of decibels above a microvolt, or dBuV, or in terms of decibels above a millivolt, or dBmV, with 0 dBmV as the 1 millivolt reference level (1 mV is 1000 microvolts). The relationship of microvolts to dBmV is listed in Fig. 2-25.

This table can be used to compare antenna specifications and for determining the amount of antenna gain required. For example, if to obtain a —6 dBmV signal (500 microvolts) at the antenna terminal when an existing antenna delivers only —20 dBmV (100 microvolts), it is indicated that a new antenna should have 14 dB of gain to increase the signal level to —6 dBmV (20 —14 = 6). You may never need to refer to this table. But it can be invaluable when you want to compare apples with apples (actual dB gain of an antenna) instead of relying on such superlatives as high gain and superhigh gain.

ANTENNA CONSTRUCTION

Antenna manufacturers, in their advertisements, specify features whose importance is seldom understood by the

(REFERENCE LEVEL: 0dBmV· 1000μv 1mv)					
dBmV	μV	dBmV	μV	dBmV	μV
−40	10.00	0	1,000	41	112,200
−39	11.22	1	1,122	42	125,900
−38	12.59	2	1,259	43	141,300
−37	14.13	3	1,413	44	158,500
−36	15.85	4	1,585	45	177,800
−35	17.78	5	1,778	46	199,500
−34	19.95	6	1,995	47	223,900
−33	22.39	7	2,239	48	251,200
−32	25.12	8	2,512	49	281,800
−31	28.18	9	2,818	50	316,200
−30	31.62	10	3,162	51	354,800
−29	35.48	11	3,548	52	398,100
−28	39.81	12	3,981	53	446,700
−27	44.67	13	4,467	54	501,200
−26	50.12	14	5,012	55	562,300
−25	56.23	15	5,623	56	631,000
−24	63.10	16	6,310	57	707,900
−23	70.79	17	7,079	58	794,300
−22	79.43	18	7,943	59	891,300
−21	89.13	19	8,913	60	1,000,000
−20	100.0	20	10,000	61	1,122,000
−19	112.2	21	11,220	62	1,259,000
−18	125.9	22	12,590	63	1,413,000
−17	141.3	23	14,130	64	1,585,000
−16	158.5	24	15,850	65	1,778,000
−15	177.8	25	17,780	66	1,995,000
−14	199.5	26	19,950	67	2,239,000
−13	223.9	27	22,390	68	2,512,000
−12	251.2	28	25,120	69	2,818,000
−11	281.8	29	28,180	70	3,162,000
−10	316.2	30	31,620	71	3,548,000
− 9	354.8	31	35,480	72	3,981,000
− 8	398.1	32	39,810	73	4,467,000
− 7	446.7	33	44,670	74	5,012,000
− 6	501.2	34	50,120	75	5,623,000
− 5	562.3	35	56,230	76	6,310,000
− 4	631.0	36	63,100	77	7,079,000
− 3	707.9	37	70,790	78	7,943,000
− 2	794.3	38	79,430	79	8,913,000
− 1	891.3	39	89,130	80	10,000,000
0	1,000.0	40	100,000		

Fig. 2-25. Decibel-to-voltage level conversion.

neophyte. For example, when it is stated that an antenna has "anodized" elements, it means that the elements are plated with a material that resists corrosion and enhances appearance. The elements of an antenna may be made of seamless metal tubing or split metal tubing, the latter being used to eliminate vibration (and possible subsequent breakage) caused by wind loading. Some elements contain a solid metal rod or wooden dowel with seamless or split tubing to minimize the breakage problem.

Radio-frequency current tends to flow along the surface of an antenna element, not through the core. This is called the "skin effect." The ideal antenna would employ silver plated copper tubing since silver has very low resistance, and because copper is also a good conductor. So is aluminum which is lighter in weight. Gold is also a good conductor, but very expensive. While silver-plated or gold-plated antenna elements have advantages over plain aluminum or copper elements, these advantages do not warrant the higher cost.

Antenna advertisements often state the number of "elements" used. Often the more elements, the better the antenna. But, this is not necessarily so. It depends upon the application of the antenna and its intended environment. And in some cases, elements may have been added to increase the number that can be claimed without actually improving the performance of the antenna. Some of the elements of a directional antenna are called "driven" elements since they feed energy to the receiver through wires. The other elements are called "parasitic" elements since they either direct or reflect radio waves.

Because of their bulk when assembled, TV antennas are usually delivered unassembled in do-it-yourself kit form. Assembly instructions are nearly always furnished and assembly is seldom, if ever, difficult if the instructions are followed.

Antenna Supports

3

Since the radio waves transmitted by a television station have some of the characteristics of light, theoretically ideal reception would be achieved when the receiving antenna literally "sees" the transmitting antenna. In most cases, it would not be feasible to install the receiving antenna at a sufficient height so that it could see the transmitting antenna. However, especially at locations at considerable distances from television stations, the antenna should be supported as high as possible above the ground. In general, the higher the antenna, the more signal that is picked up and the greater the receiving range.

REQUIRED ANTENNA HEIGHT

In areas free of hills, the required height for ideal reception can be determined by using the graph shown previously in Fig. 1-13. When the calculations indicate that an astronomically high antenna is required, the required antenna height for satisfactory reception can be easily determined by noting how high the TV antennas of the neighbors are and by checking with them about their reception.

When separate UHF antennas are used, the UHF antenna is usually mounted higher than the VHF antenna unless the UHF stations to be received are quite nearby.

At some locations, better results are obtained when the antenna is not mounted as high as possible when the TV signals are received via reflected paths.

It is not the height of the antenna above the ground that is as important as the "effective antenna elevation." When a house is located on top of a 500-foot hill, for example, the effective antenna elevation is much higher than when the house

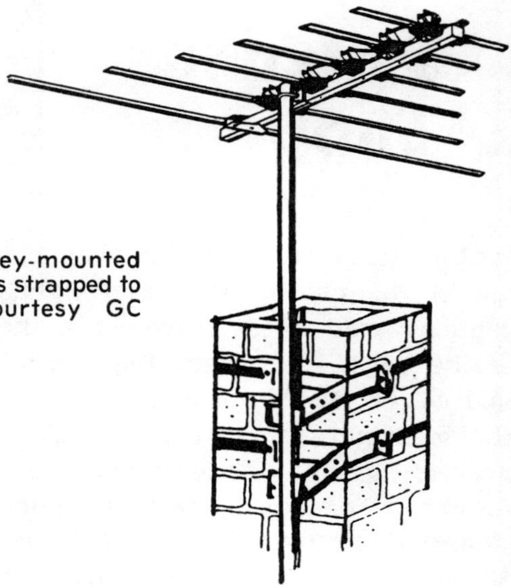

Fig. 3-1. Chimney-mounted antenna. Pipe mast is strapped to the chimney. (Courtesy GC Electronics.)

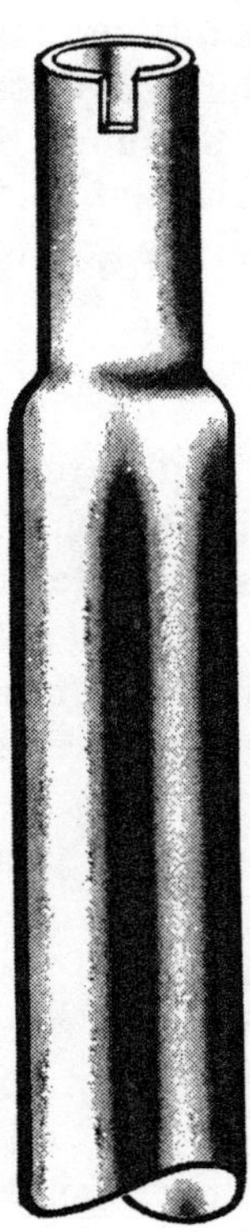

Fig. 3-2. Standard type of TV antenna mast usually available in 5- and 10-foot lengths and 1½'' and 1½'' diameters. (Courtesy Channel Master.)

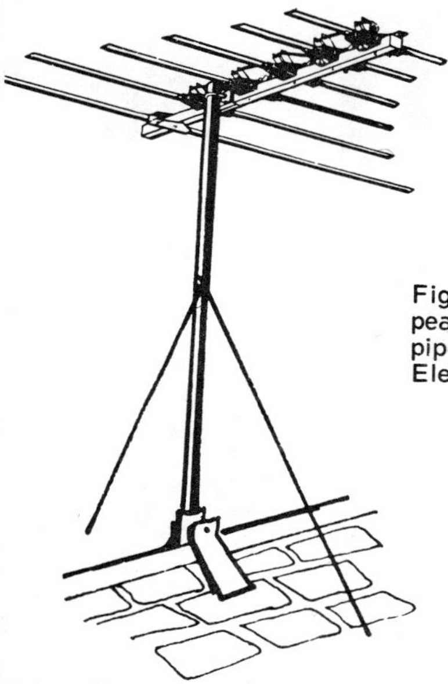

Fig. 3-3. Antenna mounted on peak of roof. Guy wires help keep pipe mast vertical. (Courtesy GC Electronics.)

is at the bottom of a valley. Similarly, the effective elevation of an antenna installed on top of a high-rise apartment building is much greater than that antenna mounted on the roof of a house only two stories high.

When an antenna extends more than 20 feet above the structure upon which it is mounted, especially in areas near airports, it may be necessary to obtain the approval of the FAA. When contemplating installing an antenna tower or other structure for supporting the TV antenna that places the antenna more than 20 feet above the roof line of the building, it is a good idea to check with the regional FAA office to determine if the installation would constitute a hazard to aircraft.

LOCATING THE ANTENNA

The antenna should be located as far away as possible from power lines and above nearby trees. Since the ignition

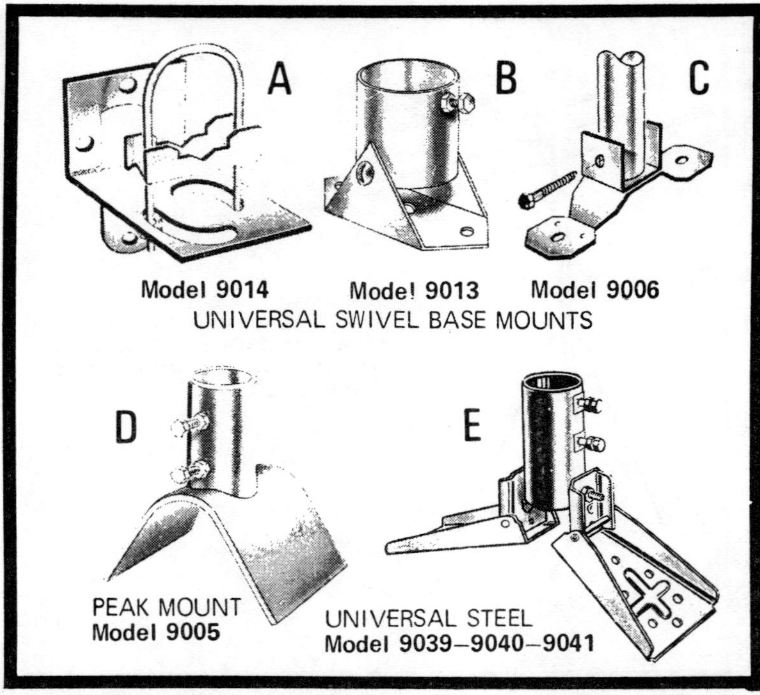

Fig. 3-4. Examples of base mounts. (Courtesy Channel Master.)

systems of automobiles and trucks generate radio interference, the antenna should be located as far as possible from the street. Also, the antenna should not be mounted close to the wall of a steel framed building which can act either as a shield or as a reflector.

CHIMNEY MOUNT

One of the easiest ways to install a television antenna is to clamp its supporting pipe mast to a chimney with chimney-mount clamps which are commonly available at hardware stores. As can be seen in Fig. 3-1, two strap mounts are used for supporting the antenna mast. It is not necessary to drill holes in the chimney. However, it should be determined that the chimney is capable of supporting the weight of the antenna and the strain placed on it during high winds.

ANTENNA MASTING

The most commonly used type of antenna mast consists of aluminum or steel pipe. An example of a standard type of TV antenna mast is shown in Fig. 3-2. This type of mast is commonly available in 5- and 10-foot lengths. As noted in the illustration, the pipe has a smaller diameter section at the top to enable stacking one length of mast on top of the other.

ROOF MOUNT

Attaching the base of the antenna to the roof of a building is another very commonly used technique. As shown in Fig. 3-3, the antenna mast is attached to what is known as a peak mount. Since this type of mount cannot support the antenna mast vertically without additional support, guy wires are used.

Examples of various types of roof mounts are shown in Fig. 3-4. The one shown in Fig. 3-4A is intended to be bolted down to a horizontal surface or a sloping surface since it is adjustable. As can be noted, it has a U-type clamp for holding the antenna mast. The mounts shown in Figs. 3-4B and 3-4C are also swivel-base mounts which are used to secure the

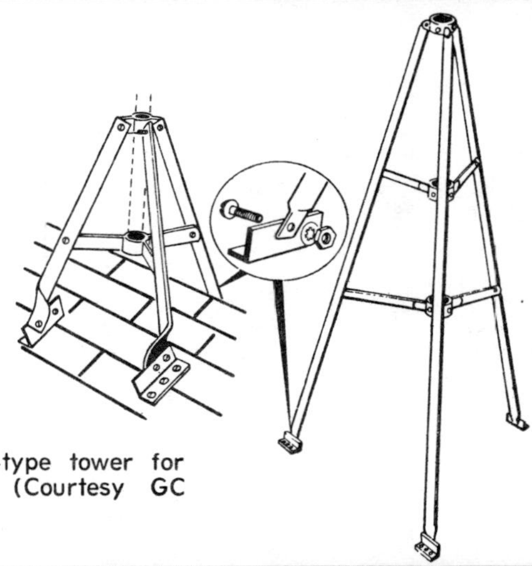

Fig. 3-5. Tripod-type tower for roof mounting. (Courtesy GC Electronics.)

A

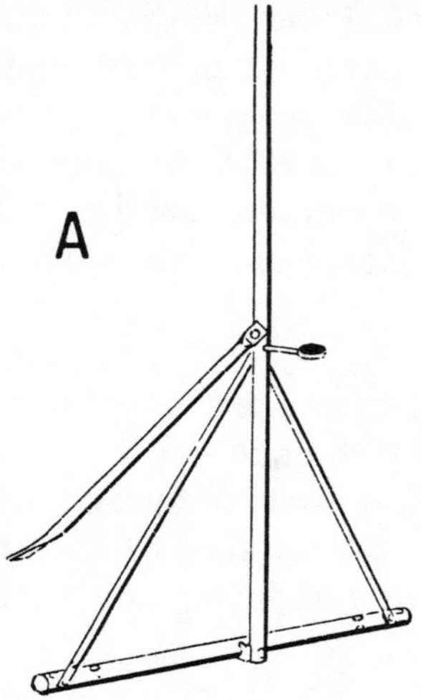

Fig. 3-6. Tripod antenna support (A) may be mounted on varying degrees of incline (B). (Courtesy Winegard Antenna Systems.)

B

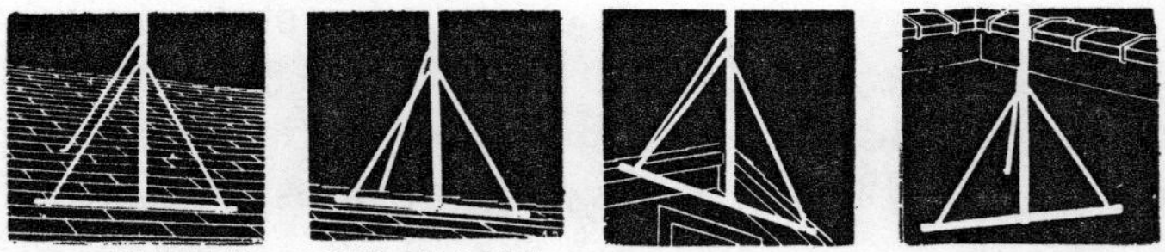

Fig. 3-7. Galvanized pipe as antenna mast.

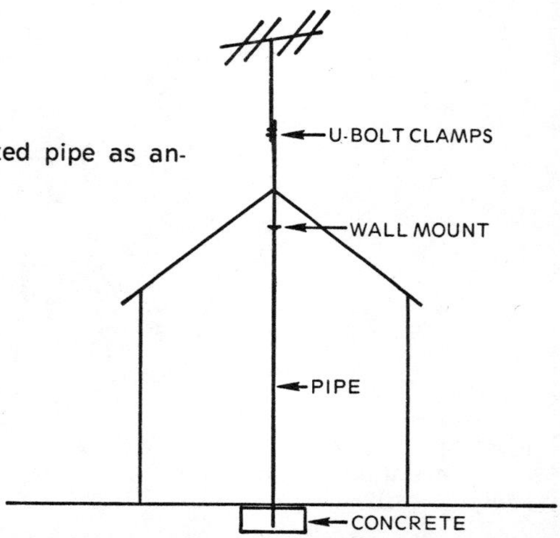

bottom of the antenna mast. The peak mount shown in Fig. 3-4D is for use where the slope of the roof is the same in both directions. The universal mount shown in Fig. 3-4E is adjustable for various angles.

TRIPOD TOWERS

Another popular type of antenna support is the tripod tower shown in Fig. 3-5. The tower is mounted at the peak of the roof in such a manner that it will hold the mast vertical. Another type of tripod tower is shown in Fig. 3-6. The tower itself is shown in (A); and in (B) is shown how the tower can be mounted on the side of the roof, top of the roof, end of the roof, or a flat roof. The advantage of the tripod tower is that guy wires are not required unless more than one mast section is used.

GROUND MOUNTED PIPE MAST

To avoid having to mount the antenna on the roof of a house, a galvanized iron pipe can be used as shown in Fig. 3-7. Here the pipe is embedded in the earth, preferably in a concrete block, and is secured to the side of the house with wall mounts. As shown in the drawing, the antenna mast is clamped to the pipe with U-bolt clamps. A U-bolt clamp assembly of this type is shown in Fig. 3-8A and a wall mount is shown in Fig. 3-8B.

INCREASING ANTENNA HEIGHT

As stated earlier, antenna mast sections can be joined to obtain increased antenna height. When the antenna support is

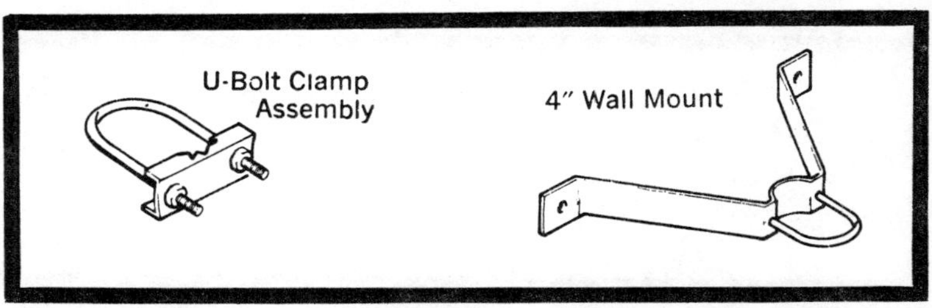

Fig. 3-8. Mast-mounting devices. (A) U-bolt clamp. (B) wall mount with U-bolt.

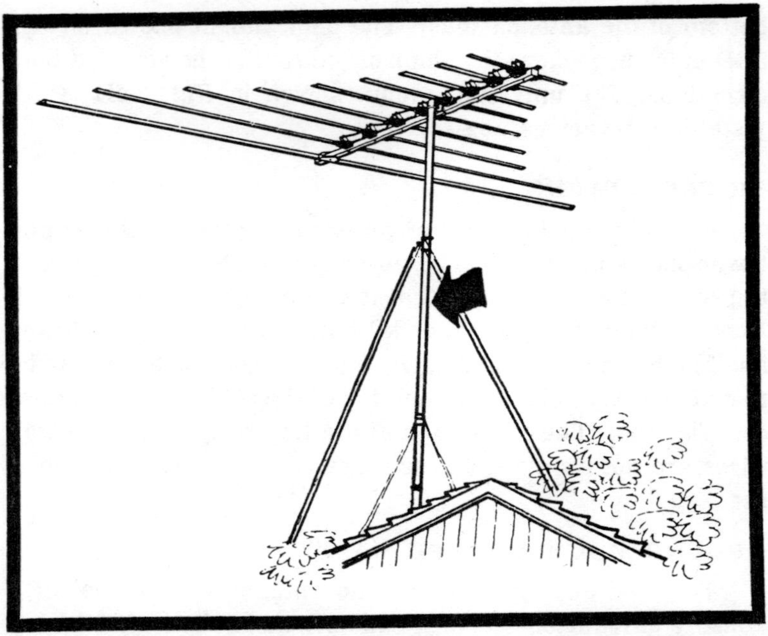

Fig. 3-9. When antenna height is increased by adding pipe section to mast, guy wires are usually needed. (Courtesy GC Electronics.)

lengthened in this manner, it is usually necessary to use one or two sets of guy wires. Figure 3-9 shows an antenna mast supported by two sets of guy wires. The guy wires can be attached to the mast by means of a guy wire clamp such as the one shown in Figs. 3-10A or by means of a guy-wire ring and collar as shown in 3-10B. The guy wire may be stranded or solid wire of sufficient strength to withstand high wind. Copper wire should not be used for guys since it can stretch under

Fig. 3-10. Guy wire fasteners. (A) mast clamp. (B) guy wire ring and collar.

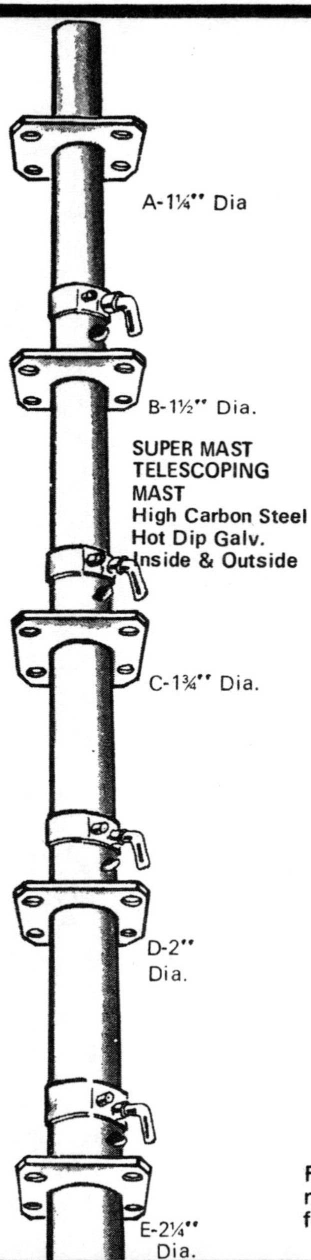

Fig. 3-11. Telescoping TV antenna mast available in lengths up to 50 feet. (Courtesy Channel Master.)

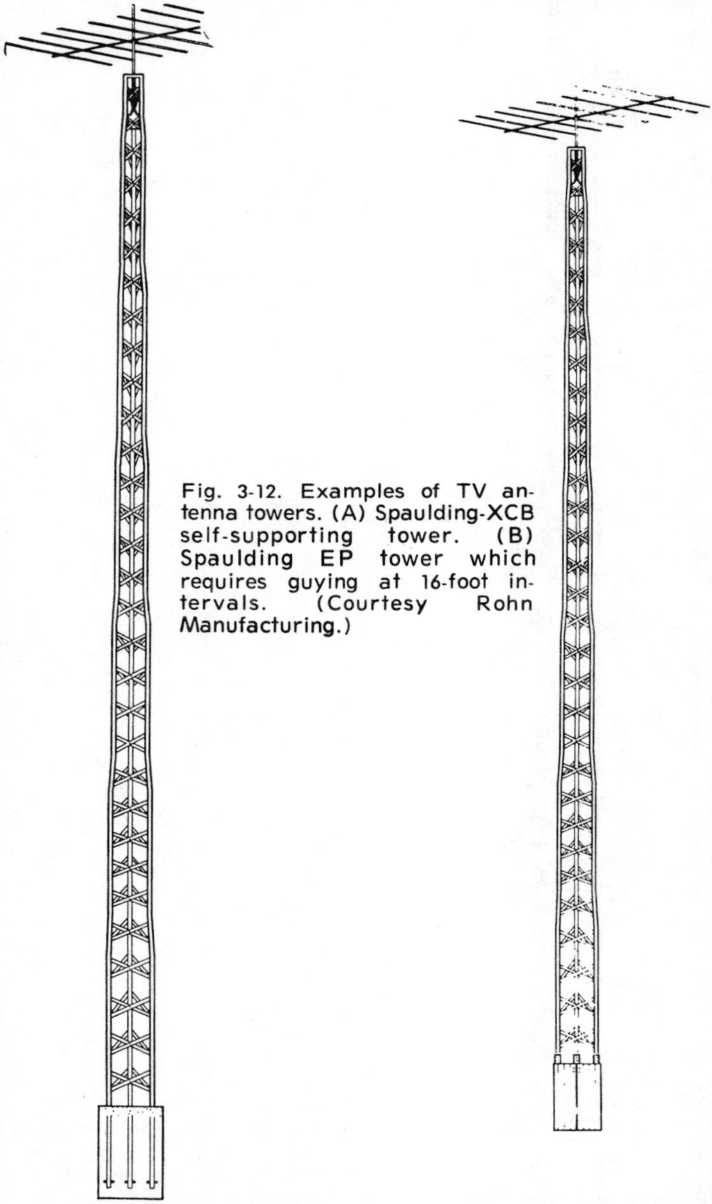

Fig. 3-12. Examples of TV antenna towers. (A) Spaulding-XCB self-supporting tower. (B) Spaulding EP tower which requires guying at 16-foot intervals. (Courtesy Rohn Manufacturing.)

tension. Radio type guy wire consists of six galvanized iron wires twisted together to form a sturdy cable. This type of guy wire is strong enough to hold masts up to 40 feet in height. The guy wires can be secured to the roof or other wooden surface by means of screw eyes. The screw eyes should be mounted at such an angle that the guy wire won't pull it out of the wooden surface. It is a good idea to install a turnbuckle in each guy wire so that the tension of each wire can be adjusted individually to keep the antenna straight and provide enough support.

TELESCOPING MASTS

When it is desirable to mount the antenna as high as 50 feet above the roof or above the ground, a telescoping TV antenna mast, such as the one shown in Fig. 3-11, can be used. As can be noted in the illustration, this particular mast consists of five sections and requires five sets of guy wires.

ANTENNA TOWERS

Steel towers, similar to those used by radio communications stations, are frequently used for supporting TV antennas where considerable antenna height is required and where there is space to install one. In some communities, it may be necessary to get a building permit or a zoning waiver to install a tower.

A self-supporting tower, such as the one shown in Fig. 3-12, takes up very little ground space. Here, the base of the tower is embedded in the earth, preferably in concrete. This tower can be assembled on the ground and hinged or built up vertically, section by section. A six-section tower of this type has an overall height of 55 feet.

A guyed tower is much less costly but requires considerable ground space for the guy wires which are required at regular intervals as shown in Fig. 3-13. It can be mounted on the ground or on the roof of a building through a suitable mount. The guy wires are usually attached to a stake driven into the ground (or to highly-secure eye bolts in the roof). The stakes should be placed approximately 8 feet from the base for

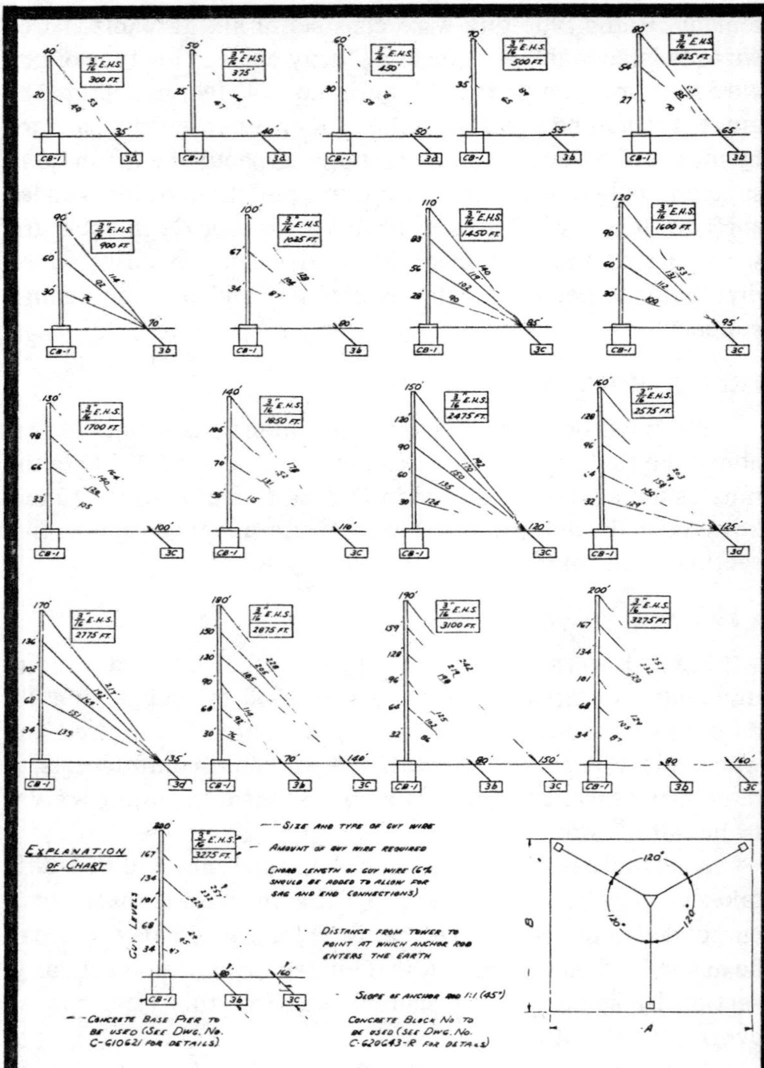

NOTES:

1) ALL GUY DIMENSIONS ASSUME LEVEL GROUND.
2) ALLOWABLE LOAD AT APEX OF TALLEST TOWER IS 180 LBS. LATERAL THRUST (EQUIVALENT AREA IS 6 SQ.FT.) DESIGN ASSUMES USE OF ONE 3/8" LINE AND ONE 1/2" LINE.
3) GUYS SHOULD BE TENSIONED TO APPROXIMATELY 10% OF THEIR ULTIMATE STRENGTH

Fig. 3-13. Guying details for typical guyed towers from 20 feet to 200 feet in height. (Courtesy Rohn Manufacturing.)

each 10 feet of tower height. For example, in the case of a 46-foot tower, the stakes should be approximately 23 feet from the base of the antenna. In lieu of guy stakes, rods, immersed in concrete, can be used. These rods should be driven into the earth at an angle so they won't be pulled straight out by the guy wires.

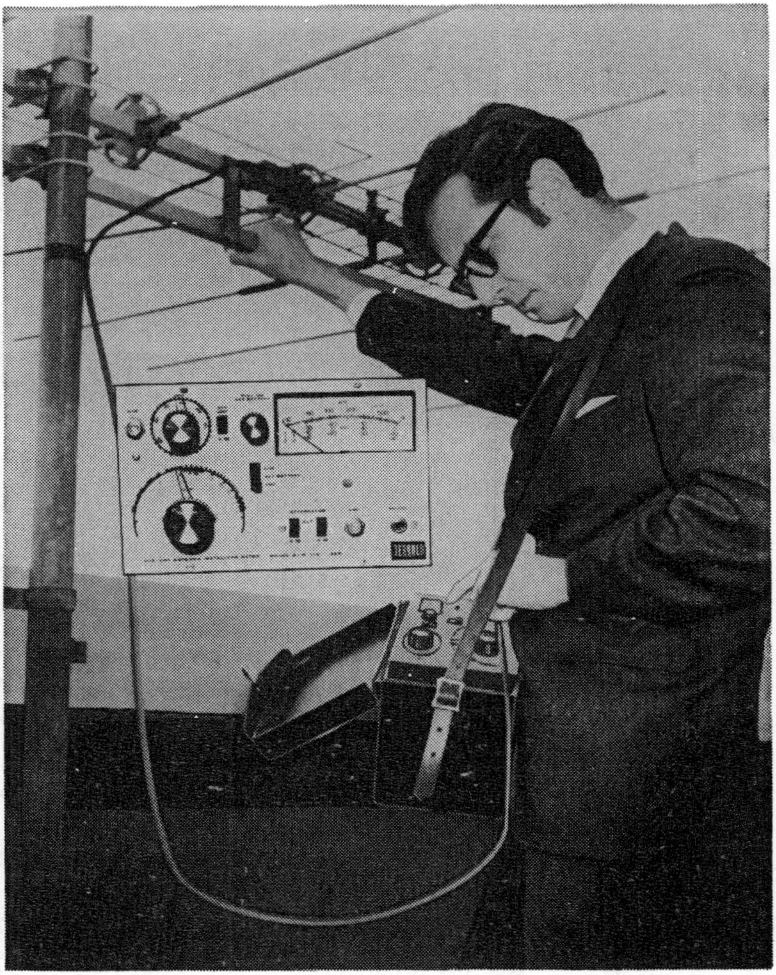

Fig. 3-14. Professional using a field strength meter to adjust antenna. (Courtesy Jerrold Corporation.)

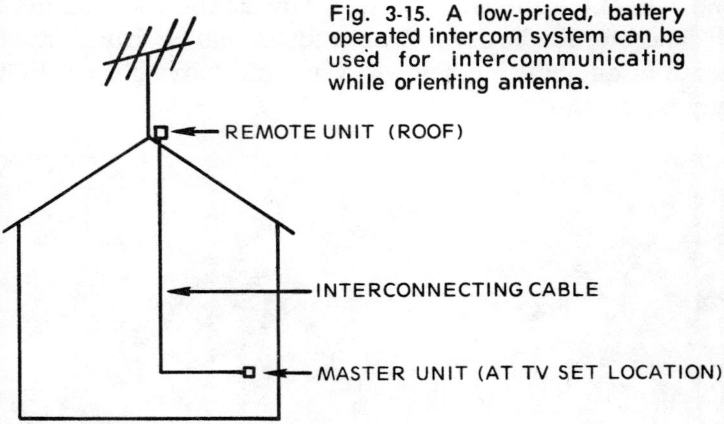

Fig. 3-15. A low-priced, battery operated intercom system can be used for intercommunicating while orienting antenna.

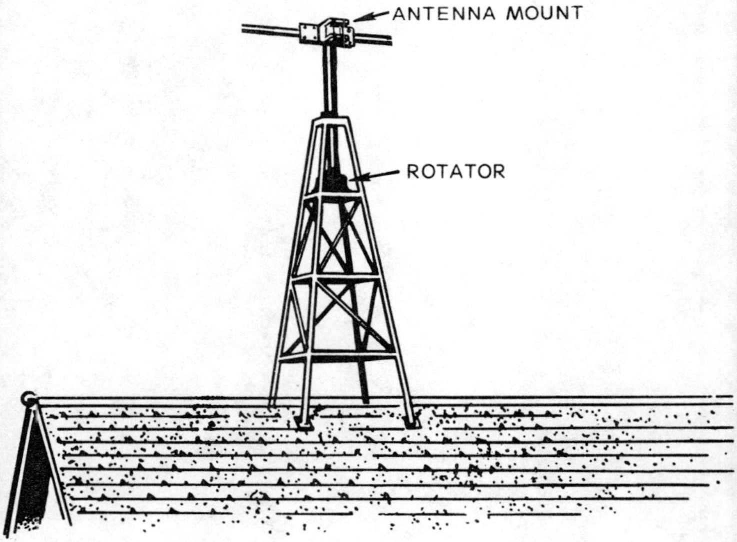

Fig. 3-16. Roof-mount tower with antenna rotator.

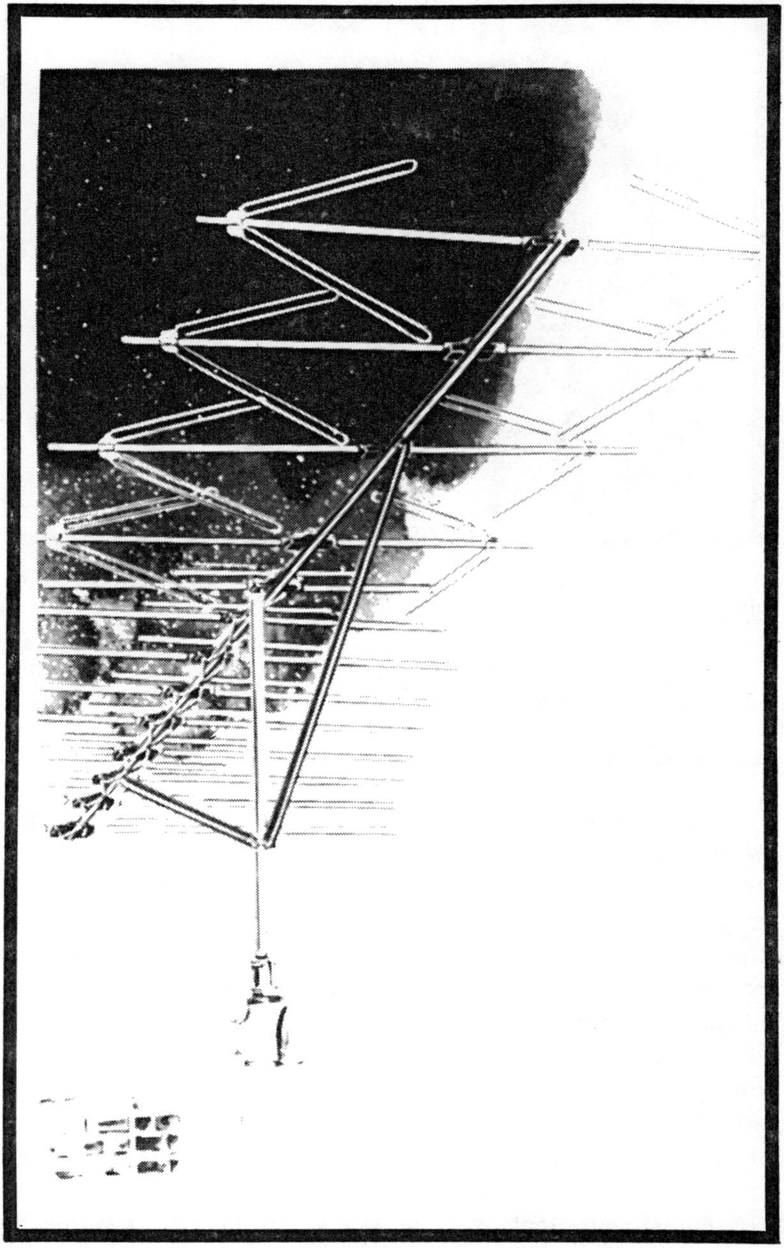

Fig. 3-17. Complex antenna attached to mast-mounted rotator.

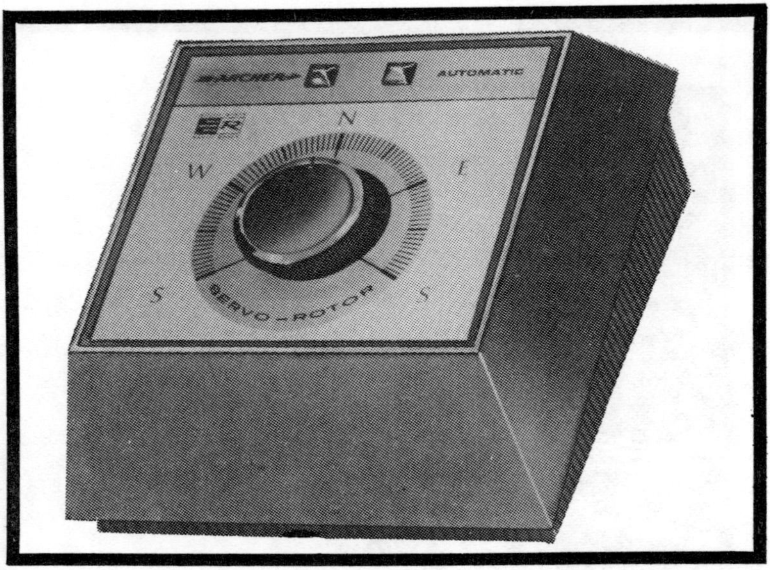

Fig. 3-18. Remote control unit for antenna rotor. (Courtesy Allied Radio Shack.)

The towers are designed to accommodate a 5- to 10-foot mast to which the antenna is attached.

ANTENNA POSITIONING

The direction in which the antenna is pointed is critical. When the antenna is mounted on a mast or tower beyond reach, the correct orientation of the antenna can be determined by first placing it on a short mast and rotating it until the clearest pictures are received. However, there may be some change in results when the antenna is moved up to its ultimate height.

A professional technician can adjust the antenna for optimum reception using a field-strength meter as shown in Fig. 3-14. The technician tunes the field-strength meter to each of the stations to be received and then adjusts the antenna for maximum indicated signal strength.

Since the typical TV receiver owner does not have access to a field-strength meter, the orientation of the antenna is usually accomplished by watching the quality of the picture of the television screen. Sometimes a portable television receiver is taken to the antenna location and connected to the antenna's transmission line and the antenna is rotated until the best pictures are received on all channels. However, this does not take into consideration the possible effects caused by the longer transmission line running down to the receiver in the home. Since it is essentially impossible for the person

Fig. 3-19. Antenna rotor mounted on a mast. (Courtesy Allied Radio Shack.)

Fig. 3-20. Antenna rotator mounted within the tower.

rotating the antenna to see the picture on the screen of the receiver in the home, it is necessary to have someone watch the picture while another person adjusts the antenna. If the distance is not too great, they can converse directly. An alternative is to use a pair of walkie-talkies at each location. Another alternative is to use a battery-operated intercom system as illustrated in Fig. 3-15. The master unit can be kept at the receiver location and the remote unit taken up to the antenna location. The two units are interconnected through a thin flexible cable. Such an intercom system is available for less than $10 from some radio supply houses.

ANTENNA ROTATORS

When the stations to be received are in different directions a rotator system can be installed which permits changing the direction in which the antenna points. As the receiver is switched from channel to channel, the rotator control knob is adjusted for the clearest picture.

The location of the rotator mounted within a roof-top tower is illustrated in Fig. 3-16. The antenna mast is driven by the rotator unit. Figure 3-17 shows a complex antenna attached to a mast-mounted rotator. The mast that is connected directly

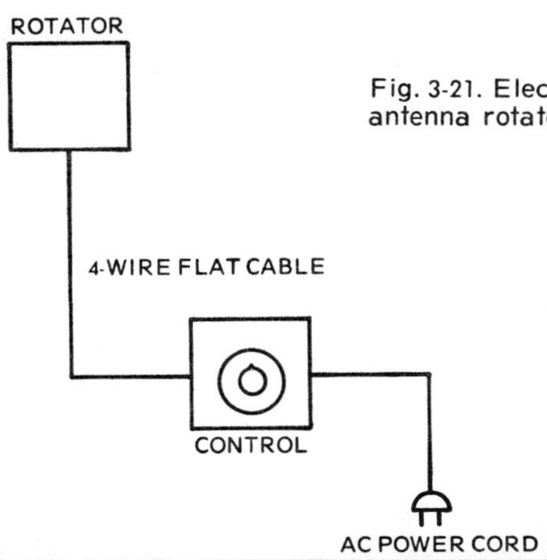

Fig. 3-21. Electrical hook-up of an antenna rotator.

to the antenna is attached directly to the rotator assembly and the rotator assembly is held by U-bolts to the supporting mast.

The rotator contains a reversible electric motor and a speed reduction assembly. The direction of rotation and its ultimate position is controlled with a remote control unit such as the one shown in Fig. 3-18.

Figure 3-19 is a closeup of an antenna rotator assembly clamped to an antenna mast. Figure 3-20 shows an antenna rotator mounted within an antenna tower assembly.

The rotator assembly is connected to the remote control unit through a 4- or 5-wire flat cable as shown in Fig. 3-21. The power cord of the remote control unit is plugged into an electric outlet. The rotator system consumes electric power only when the antenna is being rotated.

Transmission Lines

4

The transmission line is the "down-lead" or "lead-in" which connects the antenna to the TV receiver, as shown in the example illustrated in Fig. 4-1. Unlike the old single-wire lead-ins used to connect an outdoor antenna to a radio, the TV antenna transmission line must be either two-wire twin-lead or coaxial cable. (Two-wire lamp cord will not do.)

FUNCTION

The TV antenna transmission line conveys the TV signals (electrical energy) from the antenna to the TV set. To transmit this energy efficiently, the transmission line must have low losses and must be of a type designed to work with the specific type of antenna used. An electric current flows through both wires of the transmission line. These wires have "resistance" (to current flow). The longer the transmission line, the greater the resistance and the higher the loss.

When you connect a dry cell through two long wires to a flashlight lamp, as shown in Fig. 4-2, the lamp will not glow as brightly as it would if connected to the battery through shorter wires. This is because some of the electric power furnished by the battery is wasted in the wires. Because of the resistance of the wires and because of the current flowing through the wires, some of the electric power heats the wires instead of all of it being delivered to the lamp.

The same is true of a TV antenna transmission line. Look at the antenna as the battery and the TV set as the lamp. For example, if half of the power is lost in the transmission line, only half will reach the TV set. If you make the transmission line only half as long, only 25 percent of the power will be lost

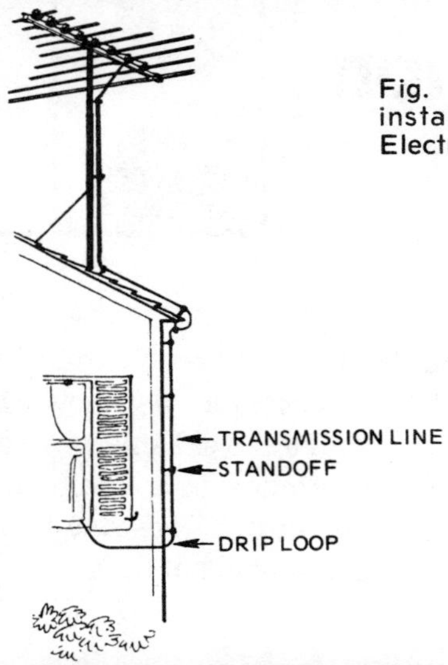

Fig. 4-1. Typical TV antenna installation. (Courtesy GC Electronics.)

and 75 percent will reach the TV set. Or if a lower-loss type of transmission line is used, more energy will reach the TV set.

TWIN-LEAD TRANSMISSION LINE

Most TV sets are designed to be connected to its antenna through what is known as a 300-ohm transmission line. This

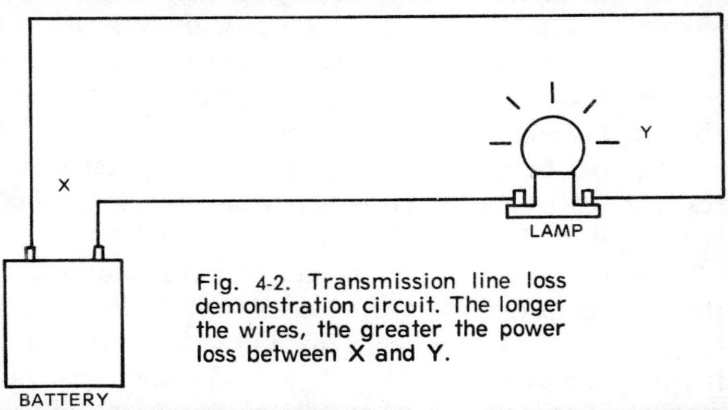

Fig. 4-2. Transmission line loss demonstration circuit. The longer the wires, the greater the power loss between X and Y.

can be any of the several types of twin-lead cable illustrated in Fig. 4-3. The most commonly used type, shown in Fig. 4-3B, is referred to as "ribbon" line.

All of these cables consist of a pair of evenly spaced wires. The "open" line consists of two solid wires spaced ½ inch apart by insulated separators at 6-inch intervals. The conventional flat twin-lead consists of two flexible copper wires embedded in a plastic material. The foam twin-lead is similar except that each conductor is covered by foamed plastic. The tubular twin-lead is round instead of flat. The shielded twin-lead is similar to foam, except that the two insulated wires are covered by a metallic shield which in turn is covered by an outer jacket.

It is the diameter of the wires, their spacing, and the characteristics of the insulating material that determine the "characteristic impedance" of the transmission line, which is typically 300 ohms. Although lamp cord consists of two flexible wires separated by and embedded in Neoprene or a plastic, its characteristic impedance is not 300 ohms and is not suitable for use as a TV transmission line.

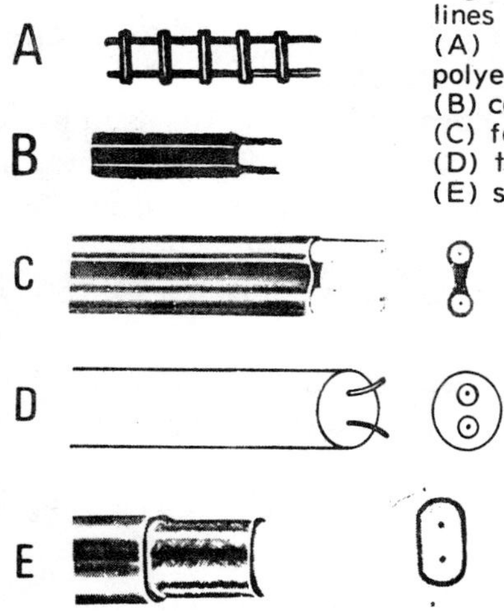

Fig. 4-3. Parallel transmission lines
(A) open wire type with polyethylene separators
(B) conventional ribbon twin-lead
(C) foam insulated twin-lead
(D) tubular twin-lead
(E) shielded twin-lead

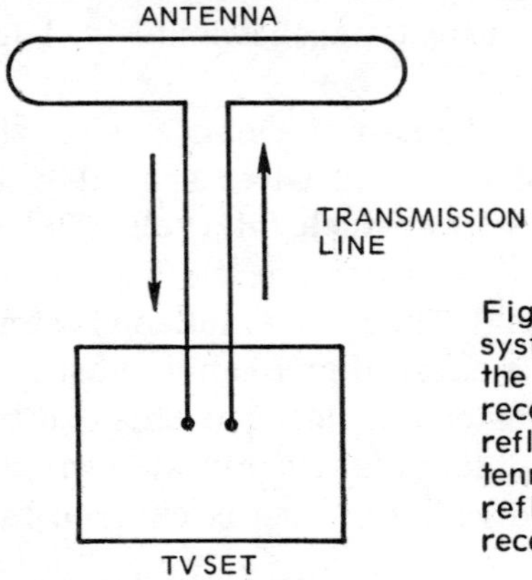

Fig. 4-4. Effects of antenna system mismatch. Signals from the antenna are fed to the receiver. A part of the energy is reflected back toward the antenna, a part of which is re-reflected back toward the receiver.

IMPEDANCE MATCHING

The output impedance of most TV antennas and the input impedance of most TV receivers is 300 ohms. To transfer maximum energy from the antenna to the receiver, the transmission line impedance should also be 300 ohms. Also, when these three impedances are the same, the transmission line itself should not pick up TV signals—it should only convey the signals from the antenna to the receiver.

If these impedances are not correctly matched, additional losses will result and the transmission line will pick up TV signals and noise. Further, ghosty pictures will result because some of the energy fed in by the antenna will be reflected back by the TV set toward the antenna, as illustrated in Fig. 4-4. This is what happens.

1. The energy from the antenna is fed to the TV set—with some loss.

2. The energy reflected back to the antenna from the TV set further reduces the amount of energy received by the TV set.

3. When the reflected energy gets back to the antenna, it is reflected back to the TV set.

4. The TV set, therefore, receives the original energy plus the energy reflected back by the antenna.

5. The direct energy travels to the TV set quickly. But the energy reflected back to the antenna and then reflected back from the antenna to the TV set takes longer to reach the TV set.

6. Since the direct energy and the reflected energy do not arrive at the TV set at the same time, the TV picture will be ghosty. The direct energy causes the wanted picture to appear on the TV set screen. The time-delayed reflected signal causes an unwanted picture to appear on the TV set screen out of register with the wanted picture.

7. In addition, the transmission line picks up TV signals which arrive at the receiver at a different time than the direct energy from the antenna—causing more ghosts.

Although the impedance of most TV antennas is rated at 300 ohms, it does not mean that this impedance is exactly 300 ohms at all TV channel frequencies. Therefore, some impedance mismatch occurs, sometimes enough to cause the transmission line to pick up TV signals and noise. This is more serious than the loss in signal strength at the receiver antenna terminals.

Even when an almost perfect impedance match is achieved, it won't stay that way when it rains or snows, when low-cost "flat" twin-lead whose characteristic impedance is affected and whose transmission losses rise when wet. All types of twin-lead (except shielded twin-lead) are affected by nearby metal. For this reason, twin-lead should never be taped to the antenna mast, a rain gutter drain pipe, or any other metal object. It should be held away from the mast, pipes, and outer walls by standoffs.

To minimize noise and unwanted TV signal pickup, the twin-lead should be twisted at intervals of less than 2 feet to form what is known as a "transposed" transmission line. This technique is used by telephone companies and railroads. The two parallel wires of a telephone circuit are criss-crossed at the crossarms of the supporting poles. When longitudinal signal current flows from east to west in the left wire, for example, between pole A and pole B, signal current flows from west to east in the right wire. If the wires are transposed at pole B, the right wire now carries signal current from east to west and the left wire from west to east. When the wires are again transposed, the situation is reversed.

Twisting twin-lead and open line has the same effect. It reduces loss of signal by "radiation" from the wires and minimizes pickup of noise and extraneous signals, including interfering signals from CB (citizens band) and amateur radio transmitters.

Long before TV came into existence and before twin-lead cable was manufactured, shortwave radio listeners and amateur radio operators used twisted lamp cord as the transmission line between a dipole antenna and the receiver or radio transmitter. Since the transmission losses are very high, twisted lamp cord is now seldom used for this purpose since superior transmission lines are now available.

SHIELDED TWIN-LEAD

To overcome the interference pickup problem as well as the effects of moisture, shielded twin-lead was developed. The two parallel wires, which are insulated from each other but which are not twisted, are surrounded by a metallic braid shield which prevents the wires from picking up noise, TV signals, and radio signals. Shielded twin-lead can be taped or otherwise attached to metal surfaces.

COAXIAL CABLE

From many standpoints, coaxial cable and shielded twin-lead are superior to any other types of TV antenna transmission line. Coaxial cable is used by telephone companies for intercity transmission of TV programs and for simultaneous transmission of hundreds of telephone conversations. Cable television (CATV) systems employ coaxial cable to transmit several TV programs simultaneously to hundreds or even thousands of homes and apartments. At TV studios, the picture signals are conveyed from the cameras to signal processing devices and amplifiers through coaxial cable. And, at both TV and FM broadcast stations, the transmitter is connected to the antenna through coaxial cable.

Coaxial cable, as shown in Fig. 4-5, is a concentric cable consisting of a solid or stranded copper or aluminum wire, known as the "inner" conductor. This wire is surrounded by solid or foam polyethylene insulation, known as the "dielectric." The dielectric in turn is surrounded by a solid or braided metal shield, known as the "outer" conductor. When a braided metal shield is used, it is surrounded by a protective plastic jacket. Coaxial cables employing a solid metal sheath-like shield may or may not have an outer protective jacket.

The characteristic impedance of coaxial cables used in TV applications is nominally 75 ohms. A 75-ohm coaxial cable may be used directly as the transmission line when the antenna's output impedance is 75 ohms and the TV receiver has a 75-ohm coaxial antenna receptacle.

However, most TV antennas have a rated output impedance of 300-ohms and most TV sets have 300-ohm antenna terminals. To use 75-ohm coaxial cable to connect a 300-ohm antenna to the 300-ohm input terminals of a TV set (or FM radio), matching transformers are required at each end of the coaxial cable, as shown in Fig. 4-6. The transformer connected between the antenna and the coaxial cable, which is known as

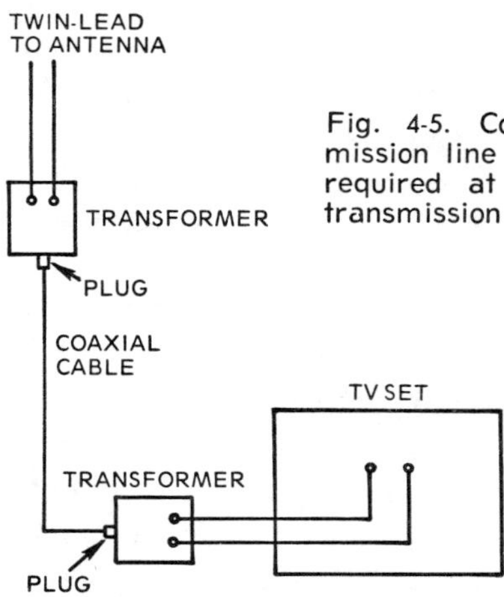

Fig. 4-5. Coaxial cable transmission line system. A balun is required at each end of the transmission line.

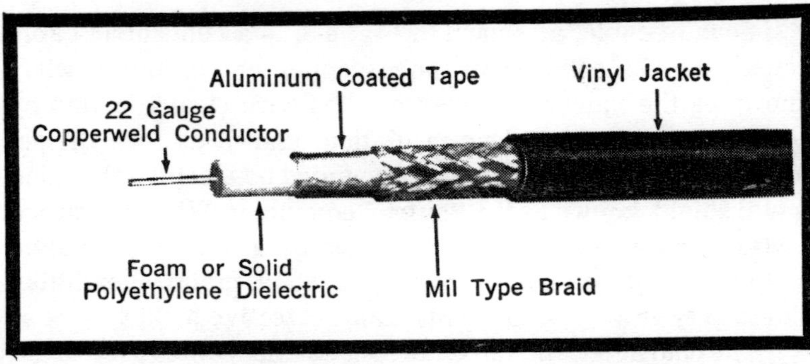

Fig. 4-6. Anti-radiation type coaxial cable which has an aluminum coated tape between the dielectric (insulation) and the metallic braid shield. Conventional type RG-59/U cable does not have this tape. Instead, the braid shield directly surrounds the dielectric. (Courtesy Vikoa, Inc.)

a "balun," matches the 300-ohm "balanced" antenna to the 75-ohm "unbalanced" coaxial cable. This transformer reduces the impedance by 4 to 1 and reduces the TV signal voltage by 2 to 1.

The transformer connected between the coaxial cable and the TV receiver, is also known as a balun. It matches the unbalanced 75-ohm coaxial cable to the balanced 300-ohm input of the receiver. It increases the impedance by 4 to 1 and steps up the TV signal voltage by 2 to 1. Since one balun reduces the voltage by 2 to 1 and the other increases the voltage by 2 to 1, there is no loss of signal voltage except due to the normal transmission losses of the coaxial cable of the small "insertion loss" of each balun.

Since its inner conductor is shielded, coaxial cable can be taped to the antenna mast, a rain gutter drain pipe, or other metal objects, standoffs are not required. In fact, coaxial cable can be fastened directly to an outer wall or any flat surface by plastic or metal clips or clamps.

Although its characteristic impedance is not affected by proximity to metal, coaxial cable should not be taped or run close to electric power wires. In spite of the fact that coaxial cable is shielded, it, nor any other transmission line, is totally immune to noise and radio and television signal pickup. But, it is far superior to twin-lead in this respect.

TRANSMISSION LINE LOSSES

Transmission line losses are not caused only by the electrical resistance of the conductors. Losses are also caused by signal leakage through the insulating material that separates the two conductors and by moisture. These losses rise with frequency; they are greater at channel 13 than at channel 2 and much greater at channel 68.

Open line has the lowest loss when dry, but increases when wet. The loss of flat twin-lead increases drastically when wet. The losses of foam-type and tubular twin-lead cable are much less affected by dampness. And the losses of shielded twin-lead and coaxial cable remain the same whether wet or dry.

Figure 4-7 shows the relative losses of all unshielded twin-lead cables with respect to shielded twin-lead. The graph

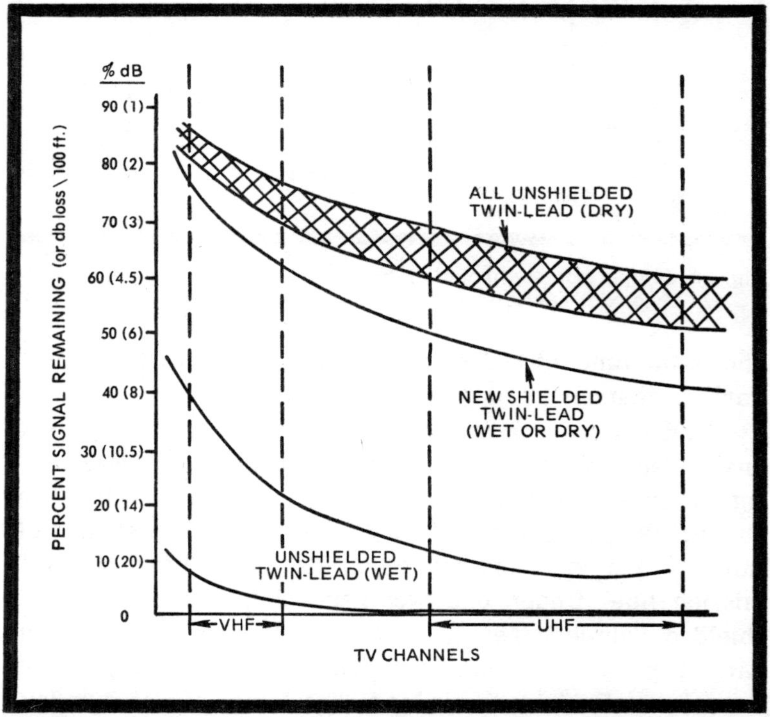

Fig. 4-7. Comparison of twin-lead and shielded twin-lead under ideal conditions. (Courtesy Belden Manufacturing Co.)

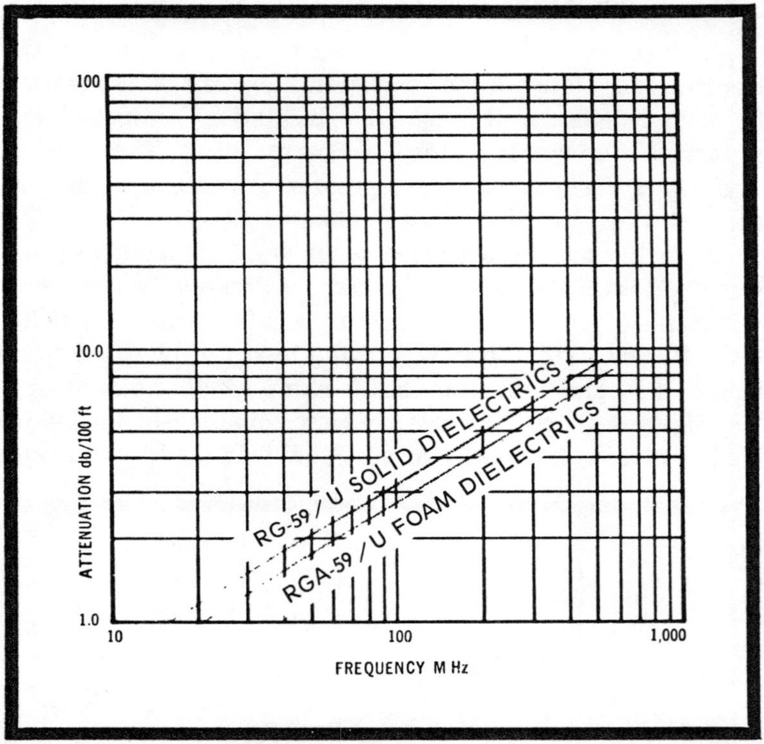

Fig. 4-8. Comparison of attenuation of solid dielectric and foam dielectric RG-59/U coaxial cable. (Courtesy Amphenol.)

shows the amount of loss of 100 feet of transmission line at various channel frequencies.

Under ideal conditions, the losses of various types of unshielded twin-lead range from about 2 dB (80 percent of signal remaining) to 3 dB (70 percent of signal remaining) at channel 13 and the losses of shielded twin-lead are about 4 dB (about 62 percent of signal remaining) per 100 feet of transmission line. Under bad weather condition the losses of shielded twin-lead remain the same but the losses of the various types of unshielded twin-lead are considerably greater than those of shielded twin-lead. The losses of flat twin-lead are greater than 20 dB per 100 feet and more than 90 percent of the signal is lost.

The losses of commonly used types of coaxial cable are considerably higher than those of any type of twin-lead. The losses of type 59/U coaxial cable (including losses caused by the transformers) are 8 dB (40% of signal remaining per 100 feet at channel 13) compared to about 4 dB for shielded twin-lead.

The losses of conventional type 59/U coaxial cable and foam-type 59/U coaxial cable at channel 12 are compared in Fig. 4-8. This graph does not include the losses in the transformers used at each end of the cable.

The term decibel, or dB, is handy to use when determining what type and length of transmission line to select. If the losses of a 100-foot length of transmission line are 2 dB (80% of signal remaining) at a specific TV channel, for example, the losses of a 50-foot length of cable would be only 1 dB (90% of signal remaining) and the losses of a 300-foot length of cable would be 6 dB (50% of signal remaining).

If the level of the signal at the antenna output terminals at a specific channel is 1000 microvolts (0 dBmV), for example, the signal level at the TV receiver input will be about 800 microvolts (—2 dBmV) when the transmission line loss is 2 dB

cable type	loss dry	rating wet	direct signal pickup rejection rating	relative cost*
open line	1	1	fair	4
flat twin-lead	4	10	fair	1
foam twin-lead	2	8	fair	2
tubular twin-lead	3	9	fair	3
shielded twin-lead	6	6	good	5
standard 59/U coaxial	10	7	good	6
foam 59/U coaxial	9	5	good	7
Colorshield coaxial	8	4	superior	10
standard 11/U coaxial	7	3	good	8
foam 11/U coaxial	5	2	good	9

*depends upon supplier prices

Fig. 4-9. Comparison of TV antenna transmission lines.

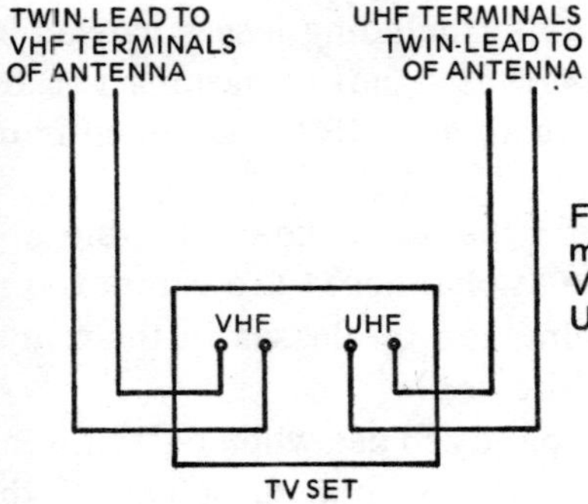

Fig. 4-10. Dual twin-lead transmission line system, one for the VHF channels, the other for the UHF channels.

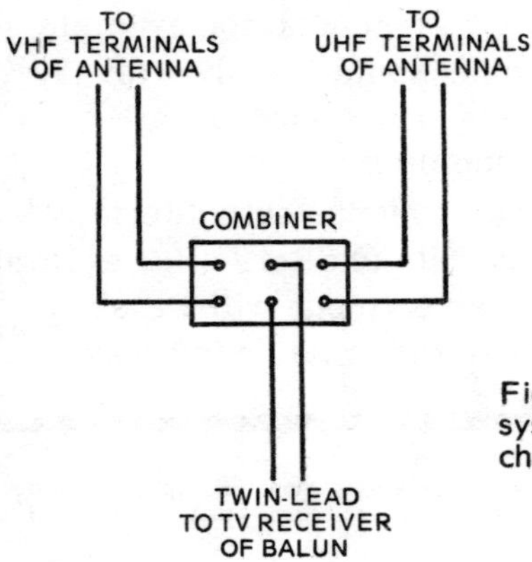

Fig. 4-11. Single transmission line system for both VHF and UHF channels.

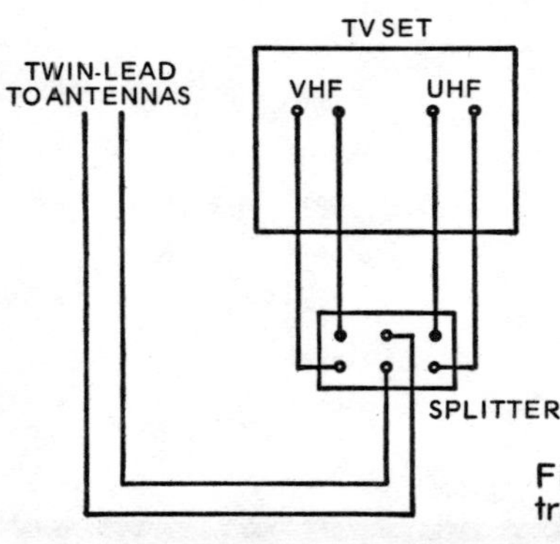

Fig. 4-12. Receiver end of a single transmission line system.

or 500 microvolts (—6 dBmV) when the loss is 6 dB. When 100 feet of conventional flat twin-lead is used whose losses are 2 dB at a specific TV channel and the losses increase to 8 dB when it is raining, the signal level will drop 50 % when raining.

TRANSMISSION LINE SELECTION

The relative merits of various types of TV antenna transmission lines are noted in the table of Fig. 4-9. Conventional flat twin-lead costs the least and can be very satisfactory in areas where it seldom rains or snows and in other areas where the TV stations to be received are nearby. Foam-type and tubular twin-lead cost more and are better. Shielded twin-lead and coaxial cable are superior because they are not affected significantly by weather and resist direct pickup of TV signals and noise, but if their losses are excessive, antenna amplifiers can be used to offset the losses.

CONNECTIONS TO AN ANTENNA

Single-channel, low VHF band, high VHF band, all-channel VHF band and an all-channel UHF band antennas

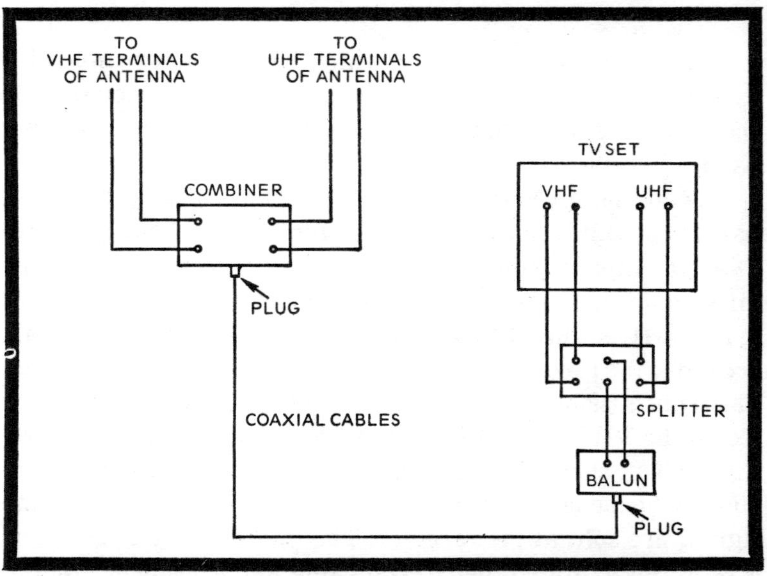

Fig. 4-13. Single coaxial cable transmission line system.

type of antenna	interface device	twin-lead connections
low band VHF	none	direct to antenna terminals
high band VHF	none	same as above
Single-channel antenna	none	same as above
All-channel VHF	none	same as above
All-channel VHF/FM	none	same as above
All-channel VHF/UHF with single pair of output terminals	none	same as above
UHF only	none	same as above
All-channel VHF/UHF with separate VHF and UHF output terminals	VHF/UHF combiner	1. VHF antenna output to combiner. 2. UHF antenna output to combiner.
Separate VHF and UHF antennas.	VHF/UHF combiner	same as above
Separate low band and high band VHF antenna.	VHF combiner	1. Low band antenna output to combiner. 2. High band antenna to combiner. 3. Transmission line to combiner output.

usually have two terminals to which the 300-ohm twin-lead is attached. (This twin-lead runs directly to the TV receiver or to the antenna-end balun when coaxial cable transmission line is used.)

Combination VHF and UHF antennas usually have two pairs of output terminals, one pair for VHF and one for UHF. The two outputs of the antenna can be fed to separate transmission lines, as shown in Fig. 4-10, or to a coupler, as shown in Fig. 4-11. Figure 4-12 shows how the VHF and UHF signals are fed to the receiver when only one transmission line is used. The transmission line is fed to a "splitter" whose outputs are fed to the VHF and UHF antenna terminals of the receiver. Figure 4-13 shows how the outputs of the antenna are combined when coaxial transmission line is used and how the signals are split at the receiver. The "balun" or transformer, is required to match the coaxial cable to the 300-ohm input of the splitter.

type of antenna	interface device	twin-lead connections
Two stacked UHF antennas.	UHF combiner	1. Output of one antenna to combiner. 2. Output of other antenna to combiner. 3. Transmission line to combiner output.
Four stacked UHF antennas	Four-input combiner	1, 2, 3, 4. Output of each antenna to combiner. 5. Transmission line to combiner output.
Two UHF antennas and a VHF antenna	UHF combiner VHF/UHF combiner	1, 2. Output of each UHF antenna to UHF combiner. 3. Output of UHF combiner to VHF/UHF combiner. 4. Output of VHF antenna to VHF/UHF combiner. 5. Transmission line to output of VHF/UHF combiner.
VHF antenna and separate FM antenna	Combiner	1. Output of VHF antenna to combiner. 2. Output of FM antenna to combiner. 3. Transmission line to combiner output.

Fig. 4-14. Antenna to twin-lead transmission line connections.

The same basic techniques can be used at the antenna end of the transmission line to combine the outputs of two VHF or two UHF antennas. In such cases, the transmission line is connected only to the receiver's VHF or UHF antenna terminals, whichever is appropriate. (See table of Fig. 4-14.)

TRANSMISSION LINE INSTALLATION

Unshielded twin-lead should be supported at intervals of 5 feet or less and held away from the antenna mast and outside of the building by standoff insulators. Examples of standoff insulators are shown in Fig. 4-15. This type is intended to be screwed into wood or through a lead plug into brick or concrete. Other types are designed to be screwed into a threaded hole in metal or are equipped with straps for attachment to a pipe.

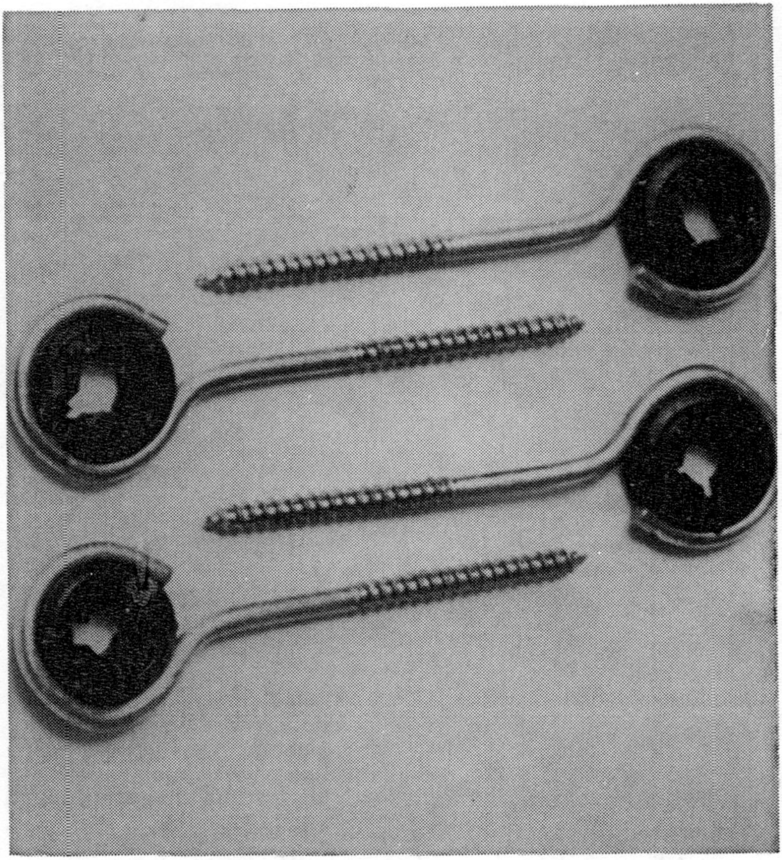

Fig. 4-15. Standoff insulators for flat or tubular twin-lead.

Shielded twin-lead can be taped directly to the antenna mast, a roof-gutter drain pipe, or held against a wall with clamps since its shield eliminates the need for standoff insulators.

One end of a shielded cable usually has terminals installed at the factory which are connected to the TV antenna terminals or antenna combiner. The jacket, shield, and drain (ground) wire usually will have been removed and a short length of shrink-fit tubing installed at a point approximately 12 inches from the terminals. If the antenna terminals are widely

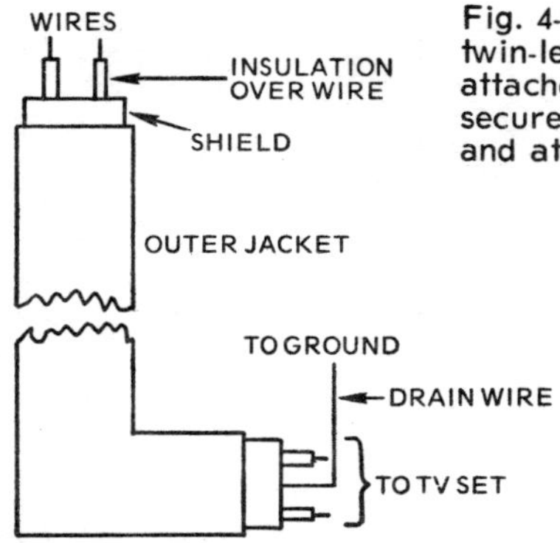

Fig. 4-16. Preparation of shielded twin-lead. Spade terminals can be attached to the wires to insure secure connections at the antenna and at the receiver.

spaced, the inner conductors can be more widely separated by splitting, or cutting the antenna end of the cable. For longer life, the antenna connections can be coated or sprayed, with a sealer such as a lacquer made for this purpose. At the receiving end of the cable, the outer jacket and shield are removed as shown in Fig. 4-16. Care should be exercised to avoid nicking or damaging the drain wire. The two inner conductors are connected to the input of the TV set. If a noise problem exists, the drain wire should be connected directly to an effective ground, as shown.

Coaxial cable can also be taped to the antenna mast or a pipe and clamped to a flat surface with cable clips or clamps (Fig. 4-17).

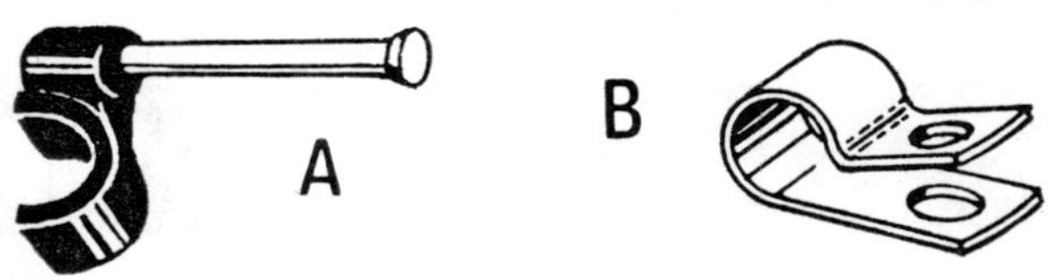

Fig. 4-17. Coaxial cable fasteners.
(A) Plastic cable clip for fastening in wood with a steel nail.
(B) Aluminum clip for fastening cable to a solid surface with a screw. (Courtesy Vikoa, Inc.)

RG-59 CABLES

ILLUSTRATION	TYPE	DESCRIPTION
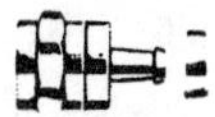	906 (F-59)	Male connector for standard 59/U type cables. "F" type.
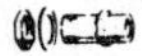	F-81	Splice, double female, "F" series.
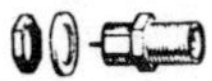	F-61	Female chassis mount, "F" series. Available in round pin or flat spring pin.
	909 (C-52)	For Standard 59/U Type Cables. "C" Type.
	(C-81)	Splice, double female, "C" series.
	(C-61)	Female chassis mount fitting, "C" series to solder point.
	918	Terminator, 75 ohms. "F" Type Connector.
	919	Terminator, 75 ohms. "C" Type Connector.
	59POT	Terminator, 75 ohms. Push-on type connector for #921 flush mounted wall plate.
	537	Adapts "C" fittings to "F" fittings.

MISCELLANEOUS

ILLUSTRATION	TYPE	DESCRIPTION
	AU-308	UHF series cable end for 308 strip braid and CF 480 size corrugated cable. Same crimping ring and tool as "F" series connectors. (Mates with SO-239).
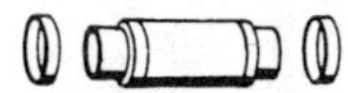	308SP	CF 480 corrugated copper cable to CF 480 corrugated cable or 308 strip braid to 308 strip braid inline splice.
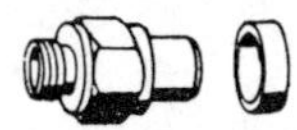	924	CF 480 corrugated copper cable or 308 strip braid to "F" type cable fitting.
	918/308	CF 480 corrugated copper cable or 308 strip braid cable female cable terminator.
	923	CF 480 corrugated copper cable or 308 strip braid cable to "F" type chassis female fitting.
	F404/201	Adapts RG11U Foam Single and Double to "F" Type Chassis Female Fitting.
	F404/59	Adapts RG11U Foam Single and Double to "F" Type Male Fitting.

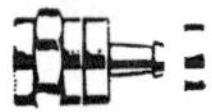	**F56**	Adapter used with #6083 UHF, VHF 75 ohm coaxial cable. Adapts down to the "F" series connectors.

INLINE SPLICE

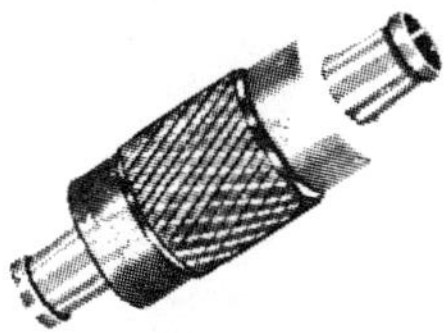	**59SP**	Inline Splice 75 ohm cables. Push-on fittings. Complete with ferrules.

FERRULES

	**#059**	For use with RG59U
	**#011**	For use with RG11U
	**#038**	For use with VK308
	**#056**	For use with F56

UHF SERIES — MIL SPEC

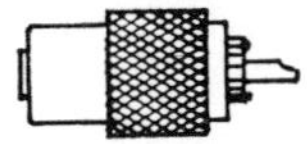	**PL259**	Male cable end Solder type.
	**PL258**	Inline splice Double female
	**M359**	90° Fitting, male/female
	**SO239**	Female chassis mount
	**M358**	"T" fitting
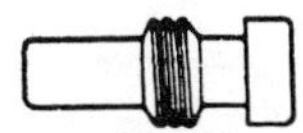	**UG175**	Adapts PL259 fitting to RG58U
	UG176	Adapts PL259 fitting to RG59U
	**DM-1**	Double male connector will accept female PL258 on both ends. Excellent for checking RF voltage at antenna output.

Fig. 4-18. Coaxial cable connectors and hardware. (Courtesy Vikoa, Inc.)

INSTALLING JERROLD SOLDERLESS CONNECTORS

PREPARING THE CABLE (CAC, CAC-6, CAD, RG-6, RG-59 TYPES)

Step 1

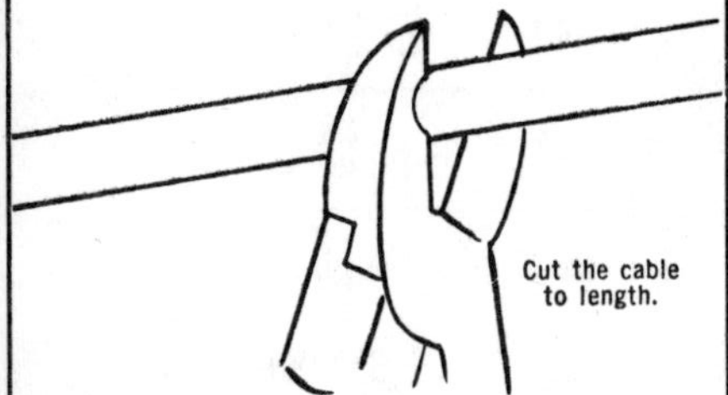

Cut the cable to length.

Step 2

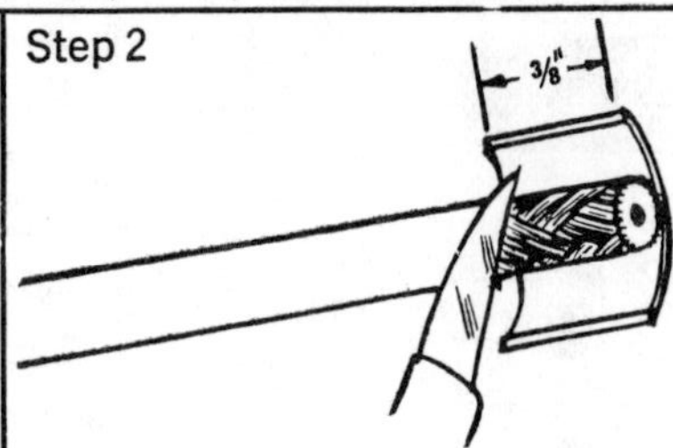

Remove 3/8" of the outer jacket with knife. Do not damage the shield.

Step 3

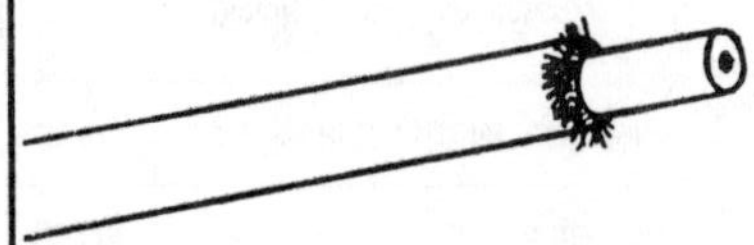

Fan the shield braid (and foil) over the outer jacket; trim close to jacket.

Step 4

Remove 1/4" of the dielectric.

Do not nick center conductor.

Step 5

Cable prepared for connector.

INSTALLING THE CONNECTOR

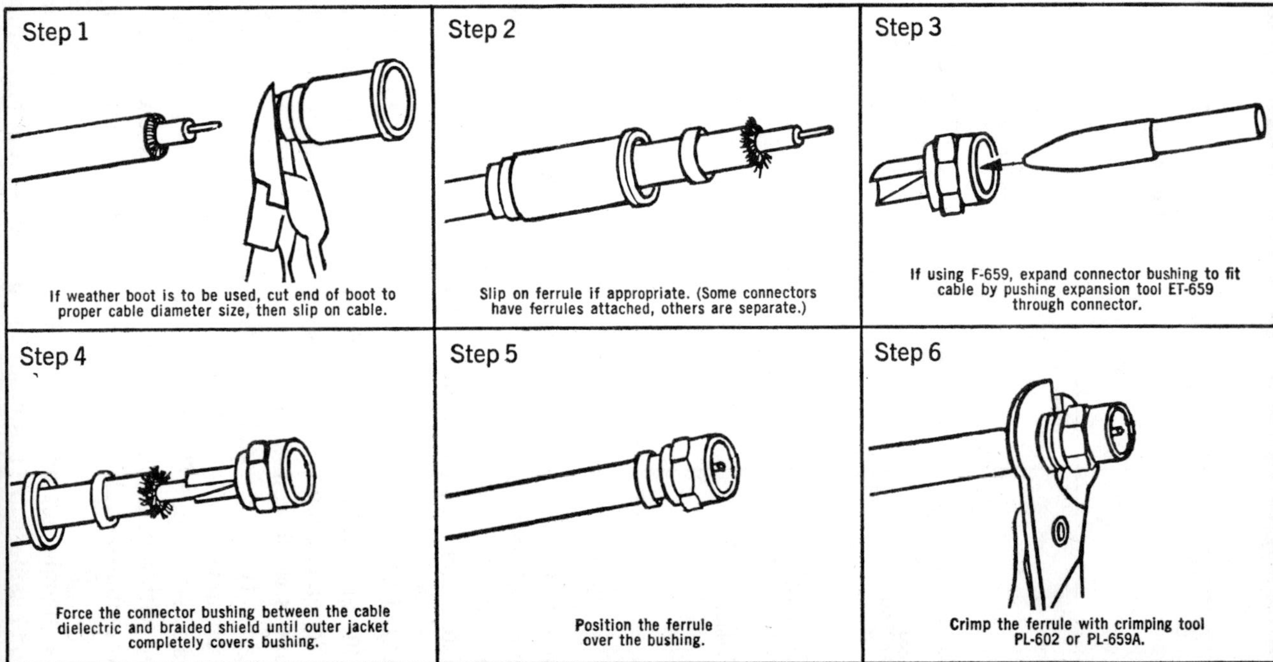

Fig. 4-19. Procedures for attaching F connectors to coaxial cable. (Courtesy Jerrold Electronics Corp.)

type of antenna	interface devices	connections
single output	balun	1. Twin-lead from antenna to balun. 2. Coaxial cable to balun.
dual output (300-ohm)	combiner balun	1, 2. Twin-lead from each antenna to combiner. 3. Twin-lead from combiner output to balun. 4. Coaxial cable to balun.
more than two outputs (300-ohm)	multiple input combiner balun	1. Twin-lead from each antenna to combiner. 2. Twin-lead from combiner output to balun. 3. Coaxial cable to balun.
single output (75-ohm)	none	1. Coaxial cable to antenna output.
two or more outputs (75-ohm)	combiner for 75 ohms	1. Coaxial cable from each antenna to combiner. 2. Coaxial cable from balun output (transmission line).

Fig. 4-20. Antenna-to-coaxial cable connections.

Sharp bends should be avoided and the cable should never be crushed. Coaxial cable should never be spliced. If a long enough piece of cable is not available, two pieces of cable can be joined together through an in-line splicing device.

COAXIAL PLUGS

A plug is attached to each end of a coaxial cable. The type of plug depends upon the type of socket on the device to which the cable is to be connected.

Various types of coaxial connectors and associated hardware are shown in Fig. 4-18. Most widely used in TV antenna systems are the type "F" and type "C" plugs. These plugs are attached with a crimping tool, such as the one shown in Fig. 4-19. Antenna-to-cable connections are tabulated in Fig. 4-20.

Most people not engaged in the TV antenna system installation business are reluctant to buy a crimping tool which they will seldom use. Type F connectors are used by CATV (cable television) and MATV (master antenna) equipment installers because one can be attached quickly to the end of a coaxial cable with a crimping tool (no soldering required) and because these connectors are inexpensive when bought in large quantities. But the typical consumer may find it difficult to buy F connectors, even at a local wholesale electronic parts distributor.

Commonly available at parts distributors and mail order houses are the widely used UHF-type PL-259 coaxial plugs which can be used with RG-11/U coaxial cable or with RG-59/U coaxial cable when using a type UG-176 adapter. The standard PL-259 plug has to be soldered to the end of a coaxial cable, as shown in Figs. 4-20 and 4-21. However, solderless equivalents of the PL-259 plug, made by Gold Line and other manufacturers, are available at parts distributors.

While the professional TV antenna installer may have F type connectors on hand, the typical consumer can more easily obtain the more ruggedly constructed PL-259 coaxial plugs. Furthermore, a better splice can be made, when necessary, by attaching a PL-259 plug to the ends of the coaxial cables to be spliced, and connecting the PL-259 plugs together through a type PL-258 adapter, as shown in Fig. 4-21. Coaxial cable should never be spliced unless absolutely necessary, and then only through connecting devices made for this purpose, as illustrated here, for example.

Although most TV receivers have only screw terminals to which 300-ohm twin-lead is connected, some do have a receptacle for connecting 75-ohm coaxial cable by means of a

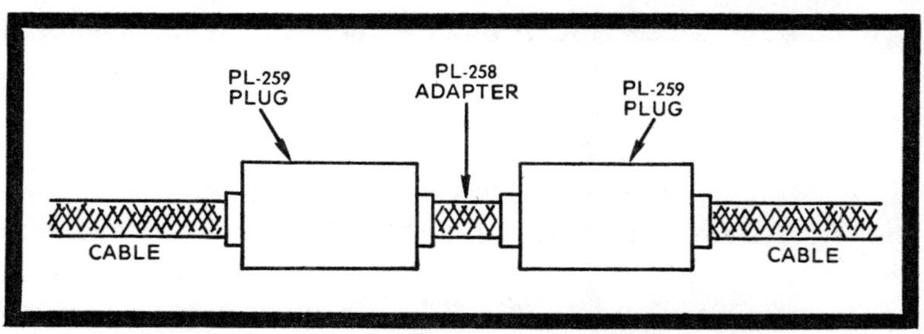

Fig. 4-21. Method of splicing 59/U or 11/U coaxial cable.

PL-259 plug or a Motorola-type auto radio antenna plug (the kind of plug used at the end of the shielded cable of an auto radio antenna). In the former case, the receiver is equipped with a type SO-239 connector which mates with a type PL-259 plug. In the latter case, an auto radio antenna type receptacle is used which accepts the Motorola-type plug. However, if the receiver has a receptacle for an F connector, an F type plug is required.

When the coaxial cable is terminated in a type PL-259 plug and the receiver has a receptacle for a Motorola-type plug, an adapter (such as Lafayette 42P22022 or equivalent) can be used to connect the transmission line to the receiver.

When the television receiver or couplers and other devices employ receptacles which require the use of F connectors, the connectors are often furnished with these devices. If not, it may be necessary to have a local electronics parts distributor order them from a CATV equipment manufacturer or supply firm (if not available locally). (The author has not been able to find them listed in mail-order house catalogs.) An alternative is to ask the local distributor to order coaxial cables of the required length with F connectors attached by the CATV supply firm. Some mail-order catalogs list RG-59/U coaxial cable in 50-, 100-, and 250-foot lengths with F connectors already attached at both ends of the cable.

When more than 50 feet of coaxial cable is required between the antenna and TV set, particularly in a weak-signal area and when an antenna amplifier is not used, the larger diameter, lower-loss RG-11/U coaxial cable should be used instead of RG-59/U coaxial cable, although it costs approximately twice as much. Since F connectors cannot be used with RG-11/U cable, type PL-259 plugs or solderless equivalents should be used.

Coaxial cables whose losses are much lower than those of RG-11/U coaxial cable often require the use of special connectors. These are usually available from the cable maker through an electronics parts distributor. When ordering special low-loss coaxial cable, it is important to state that the cable's characteristic impedance must be 72 or 75 ohms.

TRANSMISSION LINE LIFE

Twin-lead and coaxial cable deteriorate with time and may require replacement. Low-cost twin-lead might need replacement after two years of exposure to the elements. Better quality twin-lead should last longer. And shielded twin-lead and coaxial cable should last about ten years.

RUNNING CABLE THROUGH A WALL

The transmission line, whether twin-lead or coaxial cable, is usually run through a wall inside of a wall tube. It is a plastic tube which can be cut to the required length and which has bushings for each end of the tube. At the outside end of the wall tube, the transmission line should be formed to provide a drip loop to prevent water from running down the cable and into the wall tube.

Lightning Protection

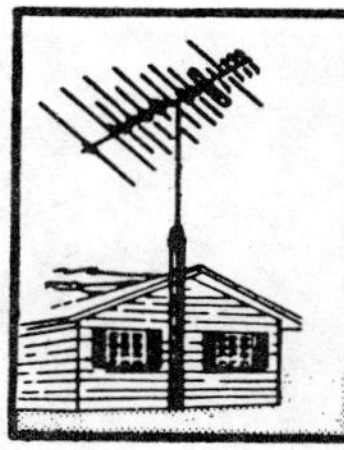

5

Statistics indicate that a television antenna does not increase the chances of a home being struck by lightning. Lightning has often struck homes without antennas. Any metal within the building, such as plumbing and electric wiring, provide a discharge path for lightning and heavy static charges. Lightning has the tendency to strike the tallest object. For example, the Empire State Building in New York has often been struck by lightning, but with no resulting damage. Since a television antenna usually extends much higher than the home, it can be struck by lightning. Figure 5-1 illustrates that the TV antenna can be affected by lightning even if it is not directly hit. A lightning bolt in the vicinity can induce a heavy current in the antenna system. The possibility of lightning damage can be greatly reduced by taking the proper lightning protection precautions.

A lightning protection system for the home is illustrated in Fig. 5-2. In this example, air terminals (lightning rods) have been installed on the roof. All of the air terminals have been connected together through a heavy wire grounded at both ends as shown in the illustration.

The television antenna mount is also grounded to the same down connector cable to which the air terminals are connected. The twin-lead from the antenna is run through a lightning arrester attached to the grounded antenna mast. In addition to the antenna mount being connected to the down conductor cable, the cable is also attached to the antenna mast as shown in the drawing.

LIGHTNING ARRESTER

Lightning arresters (Fig. 5-3) are typically simple to connect to the twin-lead transmission line. The twin-lead is

Fig. 5-1. Lightning can strike a TV antenna.

simply run through the trough in the lightning arrester and by tightening two screws, which bite into the transmission line through the insulation, both conductors of the transmission line are connected to the spark gap within the lightning arrester. While the lightning arrester can be installed near the antenna as shown in Fig. 5-2, it is more frequently connected to the transmission line at the point where the transmission line enters the building. The ground terminal of the lightning arrester can be clamped directly around a waterpipe or a ground rod as shown in Fig. 5-4.

Figure 5-5 shows how a lightning arrester can be used with shielded twin-lead transmission line. The outer protective jacket and the shield surrounding the twin-lead are removed as shown. Final tape is then applied at the beginning and end of the section of transmission line that has been stripped of its outer jacket and shield. This kind of cable contains a ground

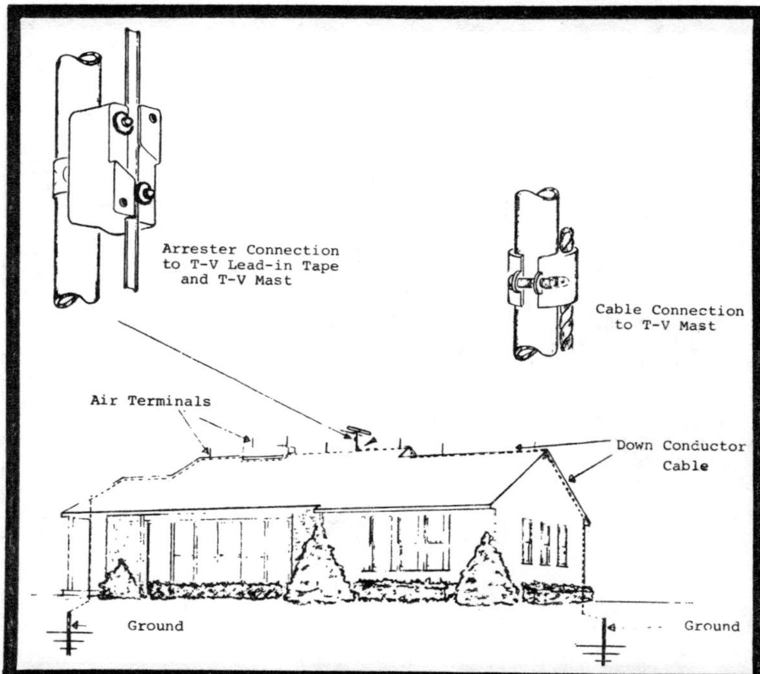

Fig. 5-2. Lightning protection for the home.

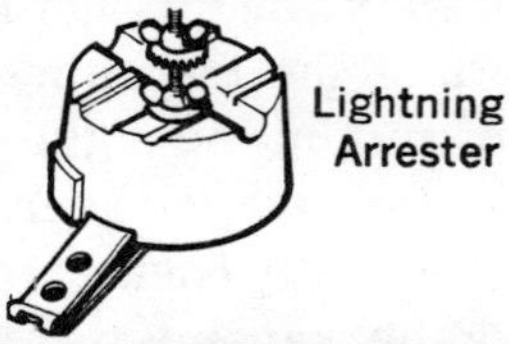

Fig. 5-3. Lightning arrester for twin-lead.

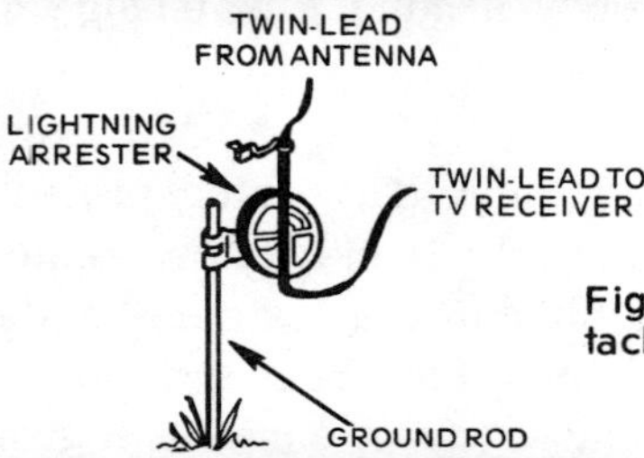

Fig. 5-4. Lightning arrester attached directly to the ground rod.

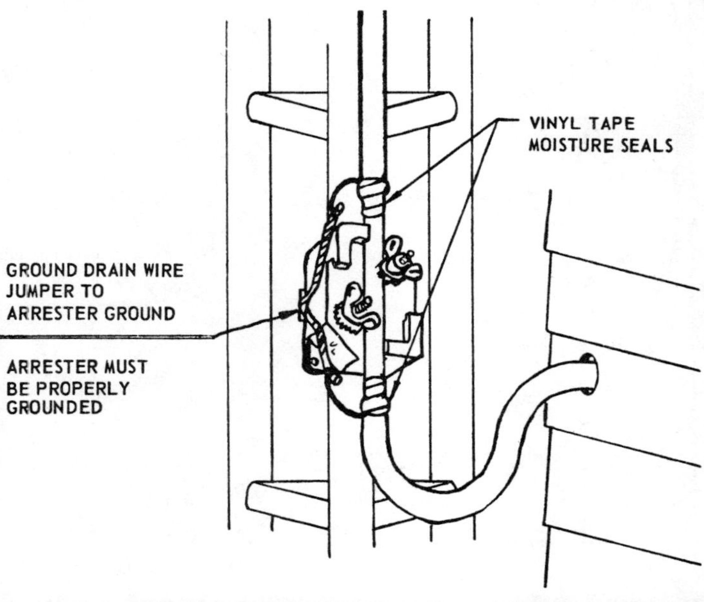

Fig. 5-5. Lightning arrester in shielded twin-lead transmission line.

drain wire which is connected to the lightning arrester ground as shown in the diagram.

When the lightning arrester is not grounded directly to a waterpipe or ground rod, it is necessary to run a heavy copper or aluminum wire from the ground terminal of the lightning arrester to a water pipe or ground rod. The connection to the pipe or ground rod should be made with an appropriate type of clamp that assures a secure electrical connection.

When a ground rod is used, it should be at least four feet long and should be driven into moist earth.

COAXIAL CABLE LIGHTNING ARRESTERS

A lightning arrester designed for use with coaxial cable can be installed at the antenna end of the transmission line. The ground connection of the lightning arrester is then securely connected to the mast which is grounded through a heavy wire. Or the lightning arrester can be installed at the point where the coaxial cable enters the building or at the balun.

GROUNDING BLOCK

One type of lightning protector for coaxial cable transmission line is the grounding block, shown in Fig. 5-6. The grounding block is screwed to the wall of the building at the point where the transmission line enters the building. The coaxial cable and ground wire are run through the grounding

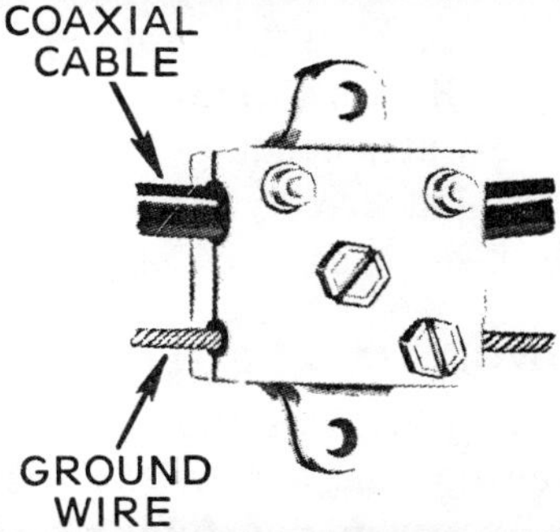

Fig. 5-6. Typical ground block.

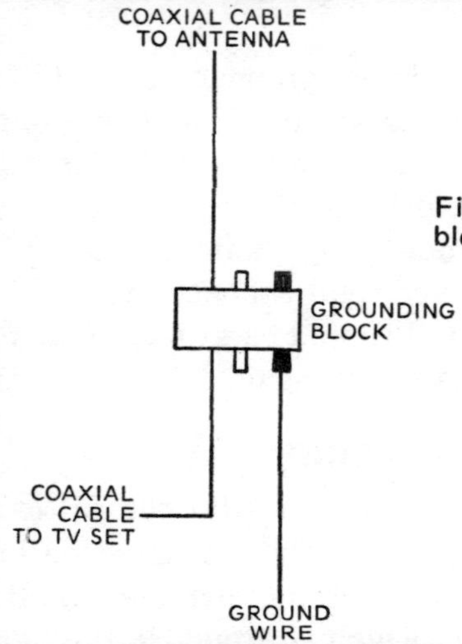

Fig. 5-7. Installation of grounding block.

block together. Figure 5-7 shows how the grounding block is wired. The grounding block is not actually a lightning arrester. However, it does provide a discharge path for heavy static charges to ground.

MAST GROUNDING

The antenna mast should be connected to the same ground wire as the air terminal. Since most homes are not equipped with air terminals to provide a high degree of lightning protection, it is a good idea to ground the antenna mast. To do so, attach a number 10 or larger copper or aluminum wire to the antenna support bolt or by means of a clamp. Run this wire as straight as possible to a ground rod or a cold-water conduit. The wire need not be insulated and can be stapled directly to the wall. It should be as short as possible and the length of horizontal runs should be minimized.

POWER LINE PROTECTION

To provide some degree of protection against damage when lightning is conveyed to a TV receiver through the power line, a power line filter can be connected between the TV

connection point	transmission line type	interface device	connections
antenna support	twin-lead	none	1. No. 10 or larger copper or aluminum wire secured to antenna support bolt. 2. Same wire to "ground" specified below
antenna support	coaxial cable	none	Same as above (optional but recommended)
Transmission line at point of entry into building	twin-lead	twin-lead lightning arrestor.	1. Twin-lead securely clamped into arrestor. 2. No. 14 or larger copper or aluminum wire from arrestor ground terminal to ground specified below.
same as above	shielded twin-lead	twin-lead lightning arrestor	1. See Fig. 5-5. 2. Same as Step 1 above 3. Same as Step 2 above 4. Drain wire of shielded twin-lead to ground terminal of arrestor.
same as above	coaxial cable	coaxial lightning arrestor	1. Coaxial cable from antenna to arrestor. 2. Coaxial cable leading to receiver to arrestor. 3. No. 14 or larger copper or aluminum wire to ground specified below.
water pipe (ground)	twin-lead	ground clamp	1. Ground wire from antenna support to ground clamp. 2. Ground wire from arrestor to ground clamp. 3. Ground wire from power line filter.
power line	twin-lead	power line filter (with ground terminal)	1. TV receiver power plug plugged into filter. 2. Filter power plug plugged into electric outlet. 3. No. 14 or larger copper or aluminum wire to ground specified below.
power line	coaxial cable	same as above	same as above

Fig. 5-8. Lightning protection connections.

connection point	transmission line type	interface device	connections
waterpipe (ground)	twin-lead	ground clamp	1. Ground wire from antenna support to ground clamp. 2. Ground wire from arrestor to ground clamp. 3. Ground wire from power line filter.
waterpipe (ground)	coaxial cable	ground clamp	same as above
ground rod*	twin-lead	ground clamp	same as above
ground rod*	coaxial cable	ground clamp	same as above

*Ground rod should be 4 feet or longer and should be driven into moist ground.
NOTE: Ground wires should be as short and straight as possible (may be stapled directly to a wall). Also separate grounds can be used for antenna support, lightning arrestor and—or power line filter to minimize ground wire length.

Fig. 5-8. con't.

receiver power cord and the electric power outlet. While such a filter is not a lightning arrester, chances are that it will ground an indirect lightning charge before it can reach the receiver. The Table in Fig. 5-8 summarizes lightning protection connecting techniques.

Interior Wiring

6

This chapter is concerned with the television signal transmission system inside of the building from the point where the transmission line enters the building. As stated earlier, the transmission line can be run into the building through a wall tube or lead-in insulator as shown in Fig. 6-1. It is important to allow a drip loop in the transmission line outside the building and to seal the outside end of the wall tube with a waterproof compound to prevent water from entering the house.

NO-HOLE ENTRY

When the homeowner does not want to drill a hole in the wall through which to run the transmission line or when an apartment dweller is not permitted to do so, the through-the-window twin-lead connector shown in Fig. 6-2 can be used. It consists of four metal discs which are cemented to the window. Two of the discs are cemented to the outside of the window and two are cemented on the inside of the window so that the inside and outside discs face each other. The twin-lead from the antenna is connected to the wing nuts of the discs on the outside of the window and the twin-lead running to the TV receiver is connected to the wing nuts of the discs on the inside of the window. These discs, while not making direct electrical contact with each other, form an electrical capacitor which will pass TV signals with very little opposition.

SURFACE INSTALLATION OF TWIN-LEAD

Where exposed to the elements, twin-lead should not be stapled or tacked to a wooden or any other type of surface. But indoors, the line can be run under a carpet or along the side of the wall. To make a neater installation, the lead can be

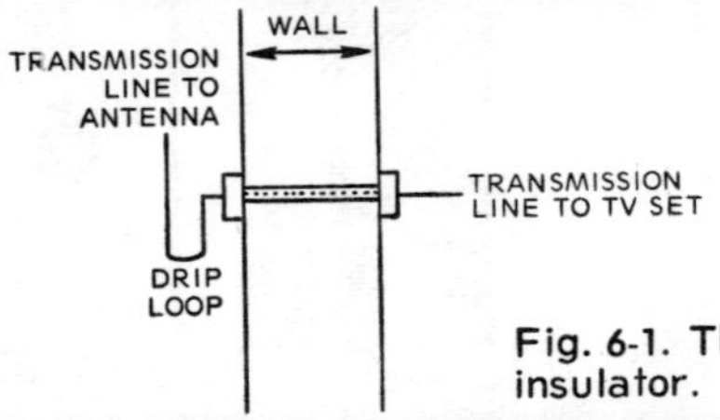

Fig. 6-1. Through the wall lead-in insulator.

fastened to the floor, molding, or wall by means of insulated nails, such as those shown in Fig. 6-3. The nail should be driven adjacent to (but not through) the twin-lead. No metallic body should be allowed between the two twin-lead wires, as metal here will change the capacitance (and characteristic impedance) of the line.

RECEIVER CONNECTIONS

When the twin-lead is used only for conveying VHF signals, the receiver end of the twin-lead can be connected directly to the VHF receiver terminals. The twin-lead is split for a distance of about 4 inches and the ends of the wires are stripped of about ⅝ inch of insulation. These wires can then be attached to the screw terminals on the back of the TV receiver. Or, to make a neater installation, lugs can be soldered to each of the wires or a solderless type of lug can be attached to each wire with a crimping tool.

An alternative is to use what is known as twin-lead clip such as the one shown in Fig. 6-4. The twin-lead is simply in-

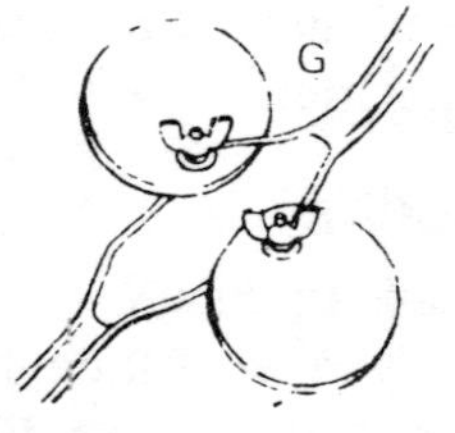

Fig. 6-2. Through-the-window twin-lead connector consists of four metal discs cemented to the window. Each pair of facing discs forms a capacitor. (Courtesy of GC Electronics.)

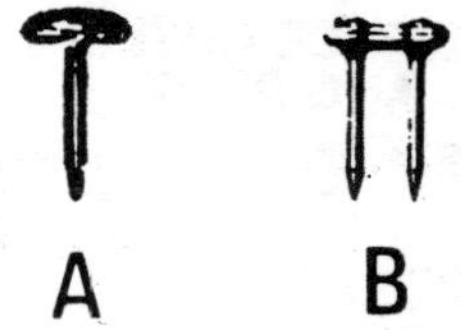

Fig. 6-3. Insulated nails for fastening twin-lead to woodwork indoors. (A) single-head. (B) double-head. (Courtesy Lafayette Radio.)

serted into the device; when the two wing nuts are tightened, the insulation is pierced and connection is made to the two wires. At the other end of the device are two stiff wires which are attached to the receiver antenna terminal.

These techniques can also be used for connecting a UHF antenna to the UHF receiver terminals.

SURFACE OUTLETS

Instead of running the twin-lead directly to the receiver antenna terminals, the twin-lead can be terminated in a wall mount socket such as those shown in Fig. 6-5. The twin-lead wires are connected to the screw terminals on the back of the socket. A short length of twin-lead is also connected to the matching plug. The far end of that twin-lead is then connected to the appropriate receiver terminals directly or through a twin-lead clip.

VHF / UHF SPLITTER

When the same transmission line is used for conveying both VHF and UHF band signals, a signal splitter is required

Fig. 6-4. Twin-lead clip. The twin-lead is attached to the clip without soldering. The two protruding clips are connected to the TV receiver antenna terminals. (Courtesy GC Electronics.)

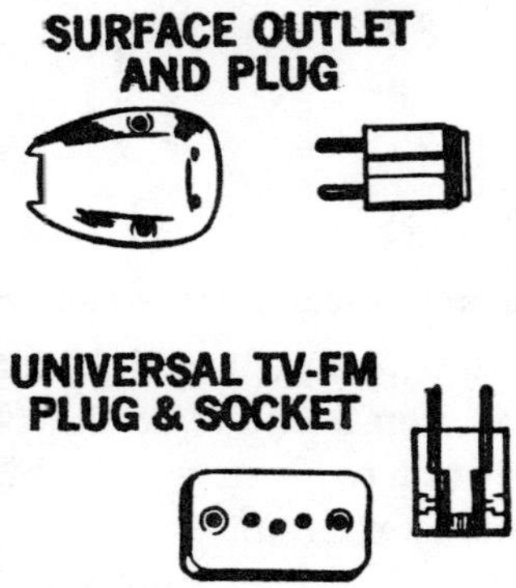

Fig. 6-5. Examples of surface-mount TV antenna outlets and plugs. Twin-lead is attached to the solderless plug by screw terminals. Courtesy Lafayette Radio Electronics Corporation.

at the receiver. The twin-lead that runs directly to the antenna system or to a wall outlet is connected to the input screw terminals of the splitter. The splitter has four output leads, two of which go to the VHF receiver terminals and the other two of which go to the UHF receiver terminals. Some splitters, such as the one shown in Fig. 6-6, also have a pair of screw terminals to which an FM radio receiver can be connected. Such a splitter channels the proper signals to the VHF, UHF, and FM radio antenna terminals.

SIGNAL DISTRIBUTION COUPLERS

A single antenna system can be used to feed signals to two or more television receivers. The twin-lead is connected to the input terminals of a coupler (actually a signal splitter) and the

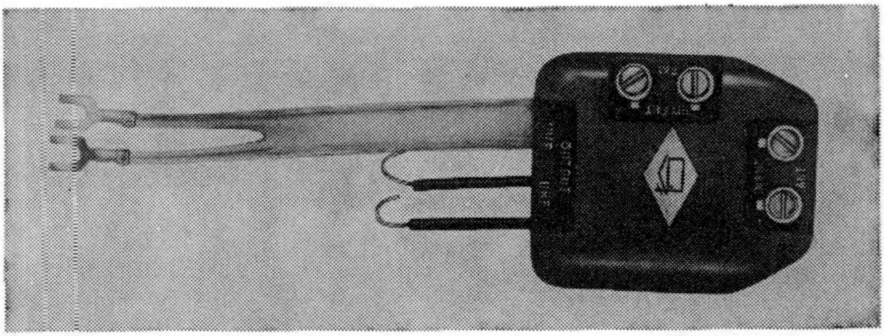

Fig. 6-6. VHF / UHF / FM splitter. (Courtesy Jerrold Electronics Corporation.)

output terminals of the coupler are connected to two or more TV receivers, depending upon the type of coupler used. Figure 6-7A shows a two-set coupler. The twin-lead is connected to the two screw terminals at the center of the coupler and the receivers are connected to the screw terminals at each end of the coupler.

With a four-set coupler (Fig. 6-7B) the twin-lead from the antenna is connected to the two screw terminals in the center and the receivers to the other terminals.

Twin-lead cables running to the different receivers should not be placed close to each other because of the possibility of mutual interference through unwanted capacitor and inductive coupling. When the antenna system is used for picking up both VHF and UHF signals, a two-set coupler can be connected as shown in Fig. 6-8. Here, the twin-lead from the antenna is connected to the input terminals of the coupler. One pair of the output terminals is connected through twin-lead to the input terminals of a VHF/UHF splitter whose output

A

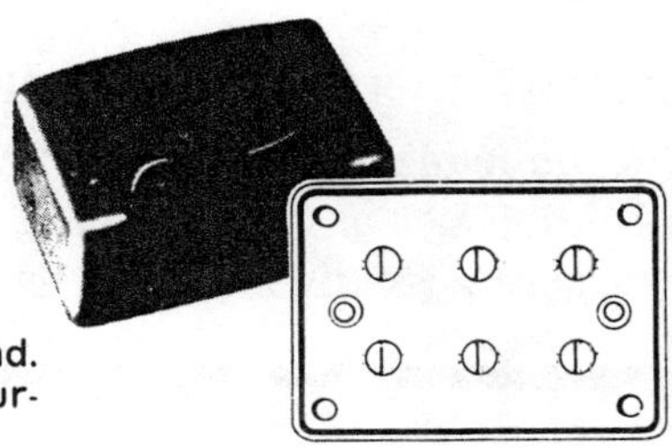

Fig. 6-7. Couplers for twin-lead. (A) two set (B) four set. (Courtesy Vikoa, Inc.)

B

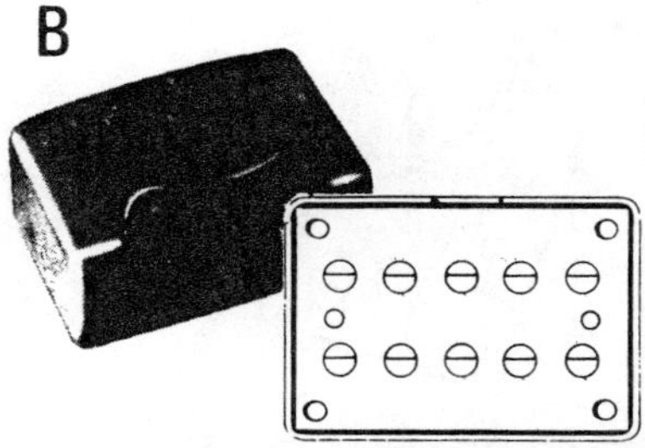

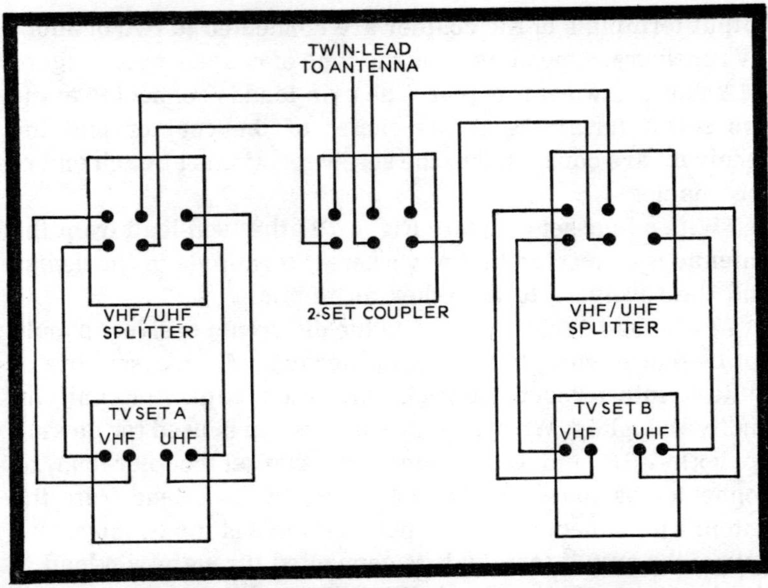

Fig. 6-8. Hook-up of two-set coupler and VHF / UHF splitters.

leads are connected to the VHF and UHF terminals of the receivers.

CONCEALED TWIN-LEAD WIRING

Twin-lead can be run within the walls and terminated in a wall outlet such as the one in Fig. 6-9. The twin-lead is secured to the screw terminals at the back of the wall outlet. Another

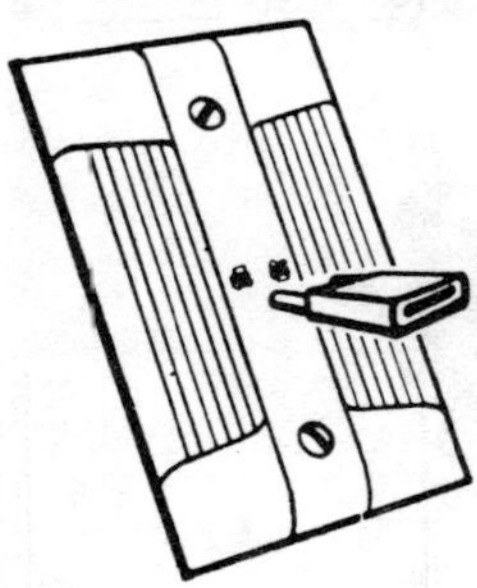

Fig. 6-9. Wall outlet and plug for twin-lead.

Fig. 6-10. Adhesive clips for supporting coaxial cable. Available for 59/U and 11/U cable. (Courtesy Winegard Antenna Systems.)

length of twin-lead that runs to the receiver is attached to the plug that mates with the wall outlet.

COAXIAL CABLE WIRING AND DISTRIBUTION

When coaxial cable is used the cable may be fastened by means of plastic or metal clips or by adhesive clips such as the one shown in Fig. 6-10. This type of clip is attached to a flat surface like an adhesive picture hanger.

When only one receiver is to be fed, the coaxial cable is fed up to the back of the set to a balun, such as the one shown in Fig. 6-11. The plug at the end of the coaxial cable is screwed into one end of the balun and the pig-tails of the balun are connected to the appropriate receiver antenna terminals.

When the same antenna system is to be used for feeding both a TV receiver and an FM receiver, a splitter with a coaxial termination can be used.

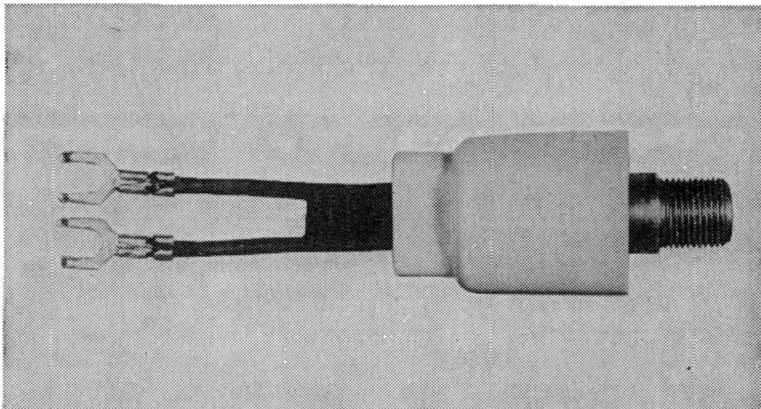

Fig. 6-11. Balun for use at TV set antenna terminals. (Courtesy Vikoa, Inc.)

Fig. 6-12. Two-way hybrid splitter for feeding two receivers or distribution lines from one transmission line, designed for connection to coaxial cable through "F" connectors. (Courtesy Electronic Switch Corporation.)

Concealed Coaxial Wiring

Coaxial cable can be run through wall to wall outlets in lieu of being surface mounted. The cable is terminated in a wall plate with a coaxial plug attached.

Wall plates with several plugs can be used for serving two TV receivers or one TV receiver and an FM receiver by using the splitter as described earlier. Splitters can be surface mount type, and can be installed in the attic or basement.

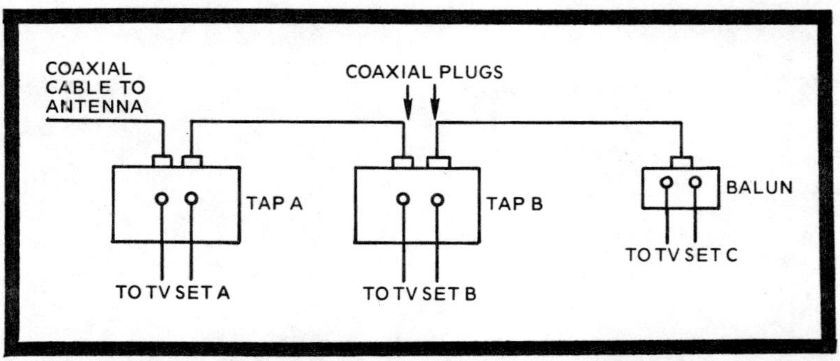

Fig. 6-13. Coaxial cable TV signal distribution system utilizing series taps.

Coaxial Cable Signal Distribution

A two-way splitter for use with coaxial cable is shown in Fig. 6-12. The coaxial cable transmission line is connected to the coaxial receptacle at the left. Each of two receivers is fed from the coaxial connectors at the right through coaxial cable and finally a balun at the receiver terminal.

COAXIAL SERIES TAPS

In lieu of using two-way VHF splitters for feeding a number of TV receivers, series-type coaxial taps can be used, as shown in Fig. 6-13. Here, the coaxial cable of the antenna is fed to the input receptacle of tap A. Its output receptacle is fed through coaxial cable to the input receptacle of tap (B) and the output receptacle of tap B is fed to a balun. These taps are provided with 300-ohm screw terminals to which the twin-lead running to the receiver can be connected.

The use of a two-way tap is illustrated in Fig. 6-14. Here the series-type taps have two coaxial tap outlets instead of screw terminals for twin-lead. As shown in the diagram, the coaxial cable from the antenna goes to the input receptacle of tap A. Its output receptacle is connected through coaxial cable to the input of tap B and its output receptacle is connected to a

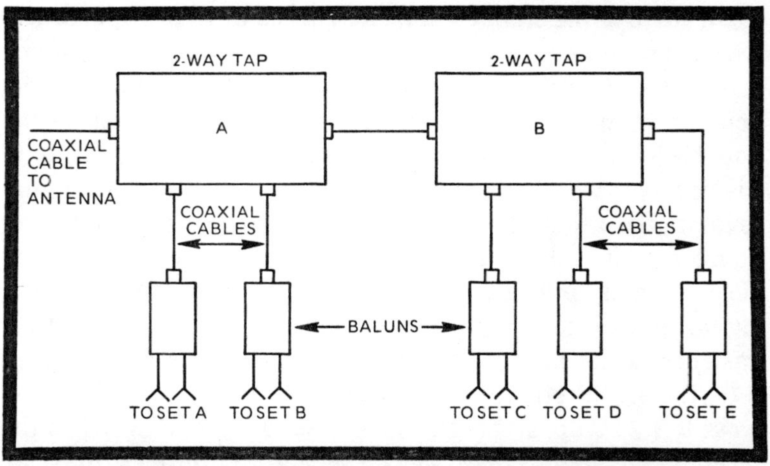

Fig. 6-14. Coaxial cable TV signal distribution system using two 2-way taps.

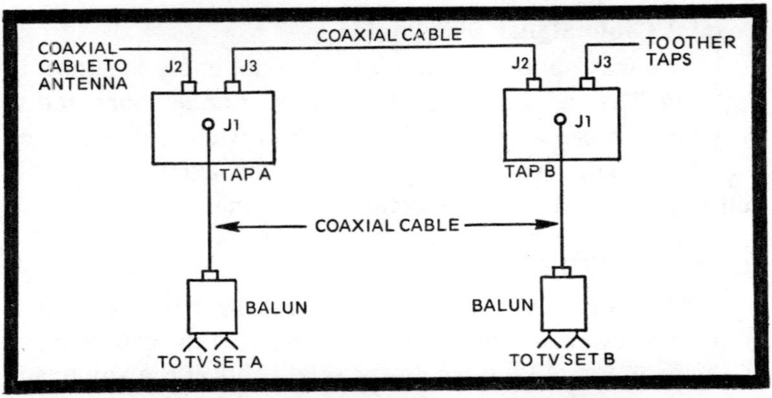

Fig. 6-15. Hook-up of wall plates with built-in taps.

balun. The baluns shown in the diagram are connected to the taps through coaxial cable and the pig-tails of the baluns are connected to the receiver antenna terminals.

NOTE: A balun is not required when the receiver has a coaxial input connector for the antenna. Most TV receivers have screw terminals for twin-lead, but some have a coaxial connector in addition to or in lieu of screw terminals.

SPLITTER AND TAP LOSSES

Whenever a coupler, splitter, series tap, or other device is used, some of the signal will be lost in the device. Supposedly, with a two-set coupler, half of the signal power is fed to one receiver and half to another. This would be totally true if the coupler did not introduce losses of its own. But it does. If the coupler loss is 1 dB for example, the signal level at the input of the receiver will be approximately 60 % as great as the signal level at the input of the coupler. In the case of a four-way coupler, the reduction in signal strength applied to each receiver will be still lower. The same is true of the various signal splitting and tapping techniques described so far. In the case of a series tap system, such as the one illustrated in Fig. 6-15, the signal strength diminishes after each tap. In areas close to TV stations, this seldom presents a problem. When the intercepted signals are weak, special amplifiers may be required as will be discussed in the following chapter.

ANTENNA SELECTION

At some locations it may be desirable to select either of two or more antennas. An ordinary double-pole, double-throw knife switch can be used, as shown in Fig. 6-16, for selecting either of two antennas. When the switch is in the position shown in the diagram, the TV set is connected to antenna B.

When coaxial cable is used at the transmission line, a coaxial switch such as the one shown in Fig. 6-17 can be used for selecting any one of up to five antennas, each fed through its own coaxial transmission line. Figure 6-18 shows how such a switch could be connected. The switch shown is a five-position type; however, there are two-, three-, and four-position coaxial switches.

CATV / EXTERNAL ANTENNA SELECTION

When a home is served by cable television (CATV), an outdoor antenna is not required. However, a TV receiver owner may want to receive channels not carried over the cable television system. A switch such as the one shown in Fig. 6-19 can be connected to the receiver to permit selection of an outdoor antenna or cable carried programs as desired. This switch has a coaxial receptacle for connection to the CATV outlet, a short length of twin-lead for connection to the VHF

Fig. 6-16. Knife switch connections for antenna selection.

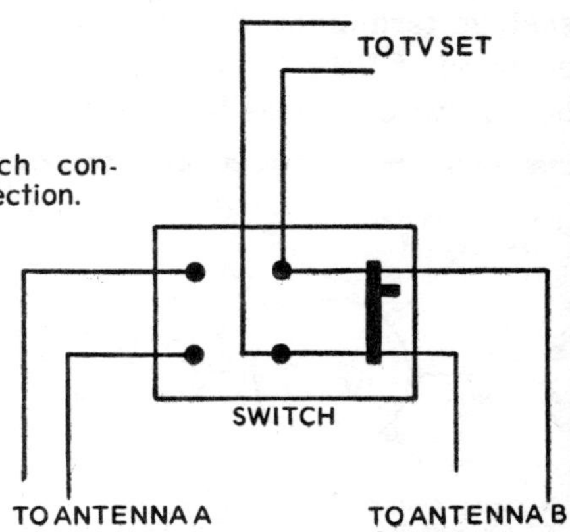

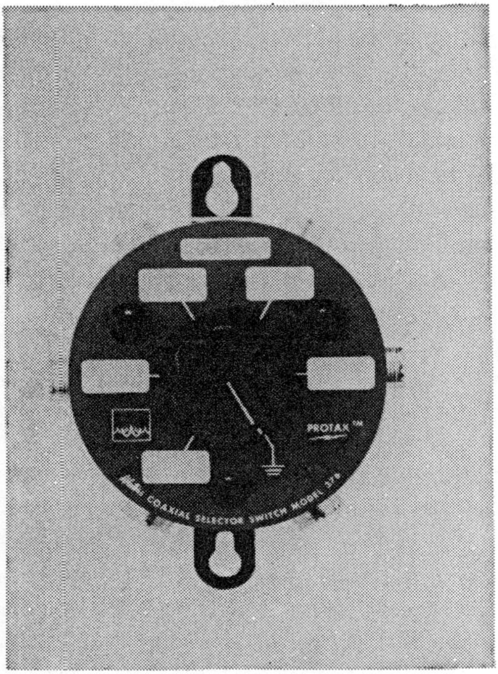

Fig. 6-17. Coaxial switch.

input of the receiver and a pair of screw terminals for connecting twin-lead leading to the outdoor antenna. Figure 6-20 shows how such a switch is wired.

Since the CATV outlet is connected only to the VHF receiver terminals, the switch does not provide for direct reception of UHF television stations. Figure 6-21 shows how such a switch would be connected to permit the use of an

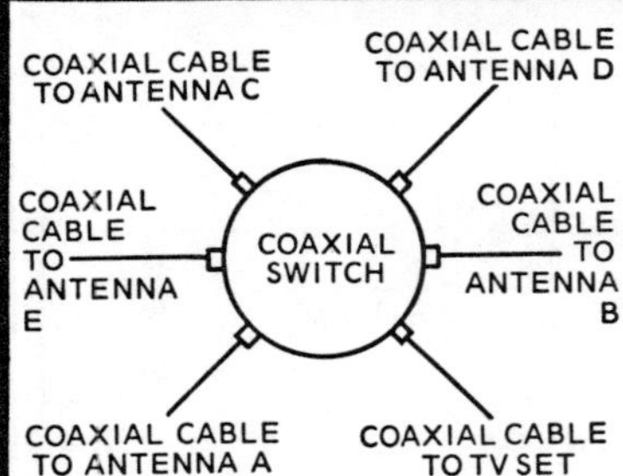

Fig. 6-18. A coaxial switch can be used for selecting any of several antennas.

Fig. 6-19. Outdoor antenna / CATV selector switch.

Fig. 6-20. Hook-up of outdoor antenna / CATV switch.

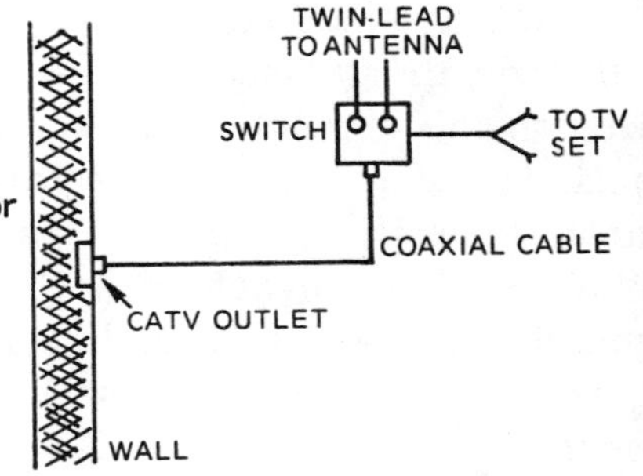

Fig. 6-21. Hook-up of outdoor antenna / CATV switch and two outdoor antennas.

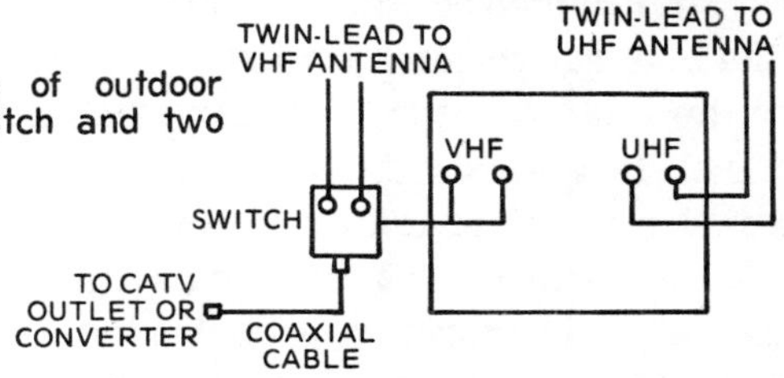

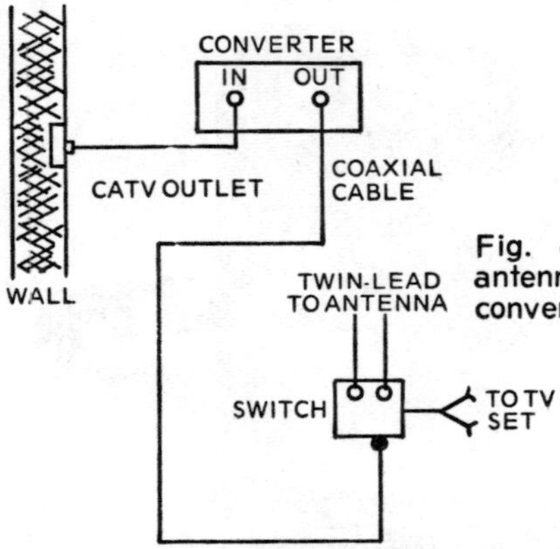

Fig. 6-22. Hook-up of outdoor antenna / CATV switch and CATV converter.

outdoor VHF antenna and how a separate UHF antenna would be connected to the receiver. The UHF antenna is connected to the receiver at all times but has no effect on the receiver except when the receiver is set for UHF reception.

CATV systems customarily convey the signals of both VHF and UHF stations. However, it is not practical to convey UHF signals through a large cable television network. To make it possible to convey the programs of UHF stations, the cable television company uses a channel translator to rebroadcast the signals of a UHF station on an otherwise unused VHF channel. The UHF station programs are received by setting the receiver channel selector to that VHF channel position.

Many cable television companies provide a converter which is connected between the CATV outlet and the television receiver. Channel selection is made with the converter instead of the channel selector switch of the receiver. In New York City, for example, the TV receiver channel selector is set to channel 12 and selection is accomplished by turning the channel selector of the converter.

To utilize an outdoor antenna with a TV receiver served by cable television through a converter, the hook-up shown in Fig. 6-22 can be used. Here, the type of switch previously shown in Fig. 6-19 is connected to the output of the converter, to the external antenna through twin-lead, and to the VHF input terminals of the TV receiver. When the switch is set for CATV reception, the receiver channel selector is set to the prescribed channel and left there; channel selection is accomplished with the channel selector switch of the converter. When the CATV switch is set into the external antenna position, the channels are selected with the channel selector of the TV receiver and the converter is cut out of the circuit. Although not shown, an external UHF antenna can be connected to the UHF terminals of the receiver.

ANTENNA SYSTEM ACCESSORIES

Most of the various TV antenna system accessories have been covered so far. The table of Fig. 6-23 is a summary of 10

type	where used	function
Antenna amplifier	antenna or near receiver	Increase strength of signal intercepted by the antenna.
Balun	antenna and receiver	Match coaxial cable to twin-lead
Combiner	antenna	Combine the outputs of two or more antennas into one composite output
Four way coupler	far end of transmission line	to enable use of same antenna by four receivers.
High-pass filter	input to TV set	Reduce or eliminate interference from radio communication transmitters
Lightning arrestor	transmission line at point of entry into building.	Protect receiver against damage by lightning and heavy static charges
Splitter	far end of transmission line	Obtain two or more outputs from one composite input signal
Two-way coupler	far end of transmission line	to enable use of same antenna by two receivers.
VHF / FM splitter	input of TV set and FM radio receiver	Separate VHF TV and FM radio signals and channel them to the appropriate receiver.
VHF / UHF splitter	receiver inputs	Separate VHF and UHF band signals and channel them to the proper receiver inputs.

Fig. 6-23. Typical antenna system accessories.

transmission line type	signals	interface devices	connections
Twin-lead	VHF	none	1. Twin-lead direct to VHF receiver terminals
Twin-lead	UHF	none	1. Twin-lead direct to UHF receiver terminals
Twin-lead	VHF/UHF	VHF/UHF splitter	1. Twin-lead from antenna to splitter input. 2. Splitter VHF pig-tail leads to VHF receiver terminals. 3. Splitter UHF pig-tail leads to UHF receiver terminals.
Twin-lead	VHF/FM	VHF/FM splitter	1. Twin-lead from antenna to splitter input. 2. Twin-lead from splitter VHF output to VHF receiver input. 3. Twin-lead from splitter FM output to FM receiver input.
Twin-lead	VHF/UHF—FM	VHF/UHF splitter VHF/FM splitter	1. Twin-lead from antenna to VHF/UHF splitter input. 2. UHF pig-tails of VHF/UHF splitter to UHF receiver terminals. 3. VHF pig-tails of VHF/UHF splitter to input of VHF/FM splitter. 4. Twin-lead from VHF output of VHF/FM splitter to VHF receiver terminals. 5. Twin-lead from FM output of VHF/FM splitter to FM receiver terminals.
Coaxial cable	VHF	balun	1. Coaxial cable to balun. 2. Balun pig-tails to VHF receiver input.
Coaxial cable	UHF	balun	Same as above except balun pig-tails to UHF receiver.
Coaxial cable	VHF/UHF	balun VHF/UHF splitter	1. Coaxial cable to balun. 2. Balun pig-tails to splitter input. 3. Splitter VHF pig-tails to VHF receiver terminals. 4. Splitter UHF pig-tails to UHF receiver input.

Fig. 6-24. Transmission line to receiver connections.

transmission line type	signals	interface devices	connections
Coaxial cable	VHF/FM	VHF/FM splitter (with 75-ohm input and 300-ohm outputs)	1. Coaxial cable to splitter input. 2. Twin-lead from splitter VHF output to VHF receiver terminals. 3. Twin-lead from splitter FM output to FM receiver terminals.
Coaxial cable	VHF/UHF/FM	balun VHF/UHF splitter VHF/FM splitter	1. Coaxial cable to balun. 2. Balun pig-tails to VHF/UHF splitter. 3. UHF pig-tails of VHF/UHF splitter to UHF receiver terminals. 4. VHF pig-tails to VHF/UHF splitter to input of VHF/FM splitter. 5. Twin-lead from VHF output of VHF/FM splitter to VHF receiver terminals. 6. Twin-lead from FM output of VHF/FM splitter to FM receiver terminals.

Fig. 6-24 con't.

transmission line type	interface devices	connections
Twin-lead	coupler with two or more outputs	1. Twin-lead from antenna to coupler input. 2. Separate twin-lead from each coupler output to VHF terminals of each receiver or to VHF—UHF splitter at each receiver.
Coaxial cable	balun coupler with two or more outputs	1. Coaxial cable to balun. 2. Balun pig-tails to input of coupler. 3. Same as Step 2 above.
Coaxial cable	75-ohm coupler balun (for each receiver)	1. Coaxial cable from antenna to coupler input. 2. Separate coaxial cable from each coupler output to balun at each receiver. 3. Balun pig-tails to VHF or FM receiver input or to input of VHF—UHF splitter whose pig-tails connect tc the VHF and UHF receiver terminals.

Fig. 6-25. Multiple output connections.

commonly used accessories including some of which will be covered in subsequent chapters. Figure 6-24 summarizes the necessary connections of the transmission line to receivers directly and through signal channeling devices. Figure 6-25 summarizes the methods of connecting couplers and baluns.

Antenna Amplifiers

7

An antenna amplifier is an electronic device that strengthens the signals intercepted by the antenna before they are fed to the input of a TV receiver. There are several types of antenna amplifiers. Some types are designed for installation at the antenna and others are designed for use at the receiver end of the antenna transmission line.

They are often called preamplifiers because they amplify the signal before it reaches the receiver.

DISTRIBUTION AMPLIFIER

A so-called distribution amplifier is one that is connected at the receiver end of the antenna transmission line and which feeds amplified signals to two or more receivers. The amplifier shown in Fig. 7-1 essentially accomplishes the same purpose as the two-set coupler previously shown in Fig. 6-9 except that the output level to each receiver is much higher than the input level from the antenna transmission line. The antenna transmission line (twin-lead) is connected to the two terminals at the right of the ac power cord. The twin-lead cables running to the receivers are connected to the four terminals at the left of the power cord. Other amplifiers of this basic type are available with more than two outputs. Some are designed for use with coaxial cable transmission lines and coaxial feeders to receivers. This particular amplifier provides 8 dB of gain. This means that the signal fed to each receiver is approximately 2.5 times as strong as the signal fed to it from the antenna.

ANTENNA MOUNTED AMPLIFIERS

The electric power required to operate the amplifier shown in Fig. 7-1 is fed directly to it from an electric power

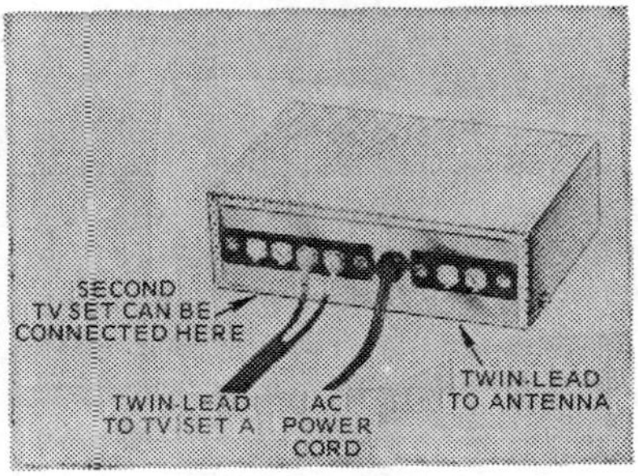

Fig. 7-1. VHF preamplifier for feeding two receivers. Provides 8 dB of gain.

outlet through its ac power cord. Electric power to an antenna mounted preamplifier, on the other hand, is fed to it through the antenna transmission line without interfering with television signals. Figure 7-2 shows the connections of an amplified antenna system in which the amplifier is located at the antenna. The twin-lead from the antenna is connected to the input terminals of the amplifier. The output of the amplifier is fed through a coaxial cable to the power supply unit near the receiver. Since the amplifier input is 300 ohms and its output is 75 ohms, the amplifier functions as an amplified balun. The signals fed into the power supply unit through the coaxial cable are fed right through the power supply unit

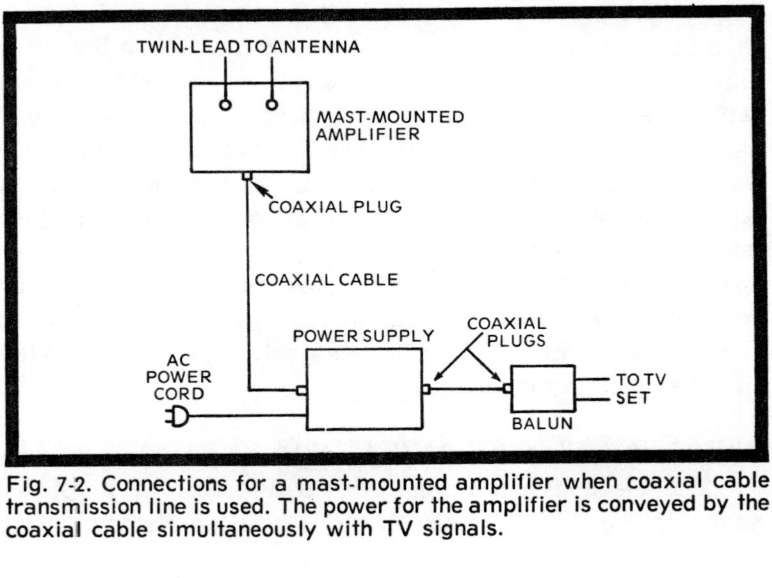

Fig. 7-2. Connections for a mast-mounted amplifier when coaxial cable transmission line is used. The power for the amplifier is conveyed by the coaxial cable simultaneously with TV signals.

through another coaxial cable to a balun whose pigtails are connected to the input terminals of the receiver. The power supply unit contains a rectifier which changes the ac voltage into dc required for operation of the amplifier. This dc power is transmitted through the coaxial cable to the amplifier simultaneously with television signals being transmitted down the coaxial cable, through the power supply and to the receiver.

Examples of other antenna mounted amplifiers are shown in Fig. 7-3. The one shown in A is designed for use with a coaxial cable transmission line. The one shown in B is designed for use with a twin-lead transmission line. The one shown in C is also designed for use with a coaxial cable transmission line, but has separate UHF and VHF antenna inputs. It therefore functions as a VHF / UHF antenna combiner as well as amplifier. The one shown in D is designed for use with twin-lead transmission line and has both UHF and VHF antenna inputs.

UHF BAND AMPLIFIER

In some cases it is desirable to amplify both VHF and UHF band signals. In other cases, it is only necessary to amplify UHF band signals. The amplifier shown in Fig. 7-4 is designed for mounting to the antenna mast and for use with a coaxial cable transmission line. As shown at the bottom, the amplifier has two coaxial connectors. The antenna is con-

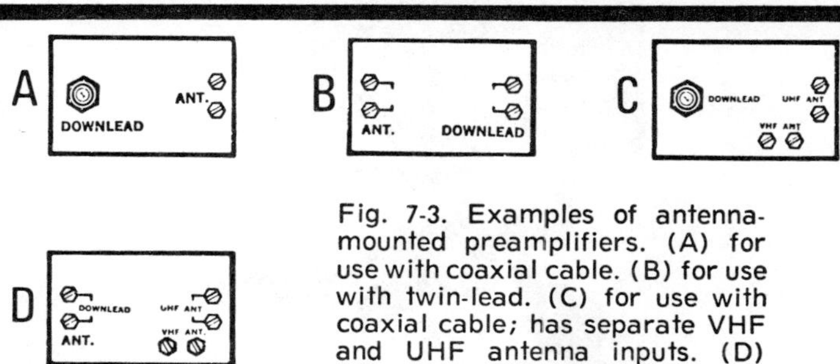

Fig. 7-3. Examples of antenna-mounted preamplifiers. (A) for use with coaxial cable. (B) for use with twin-lead. (C) for use with coaxial cable; has separate VHF and UHF antenna inputs. (D) same as (C) except for use with twin-lead.

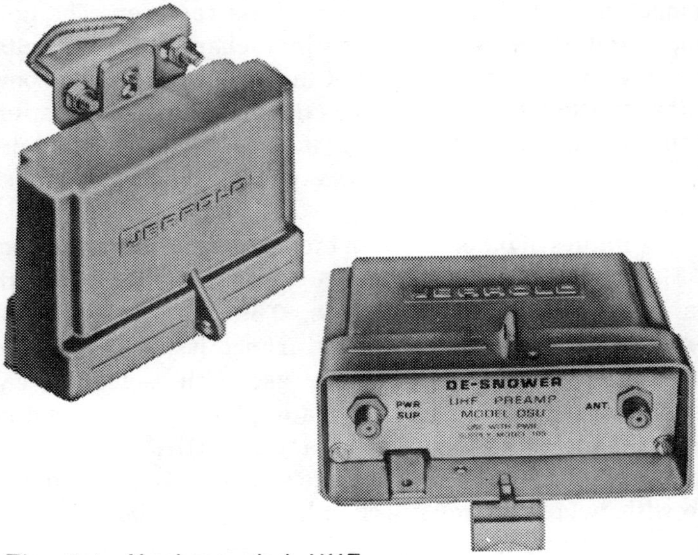

Fig. 7-4. Mast-mounted UHF preamplifier.

Fig. 7-5. Power supply for mast-mounted preamplifier.

nected to a balun and then through a short length of coaxial cable to the coaxial connector marked ANT. The coaxial cable transmission line is plugged into the coaxial connector marked PWR SUP. The electric power required to operate the amplifier is fed up to the amplifier through the coaxial cable transmission line and the UHF band television signals are fed down the transmission line to the power supply and thence to the receiver.

An example of a power supply for an antenna mounted amplifier is shown in Fig. 7-5. The coaxial cable transmission line is connected to the coaxial receptacle at the right and the coaxial cable leading to the television receiver through a balun is connected to the coaxial receptacle below it. The slide switch at the bottom left can be set in any of four positions to vary the amount of operating voltage fed up to the amplifier. If the transmission line is less than 25 feet long, the switch is set to its extreme right position. If the transmission line is between 26 and 50 feet in length, it is set to the next left position, if it is from 51 to 75 feet long, the switch is set to the next adjacent left position and if the coaxial cable transmission line is

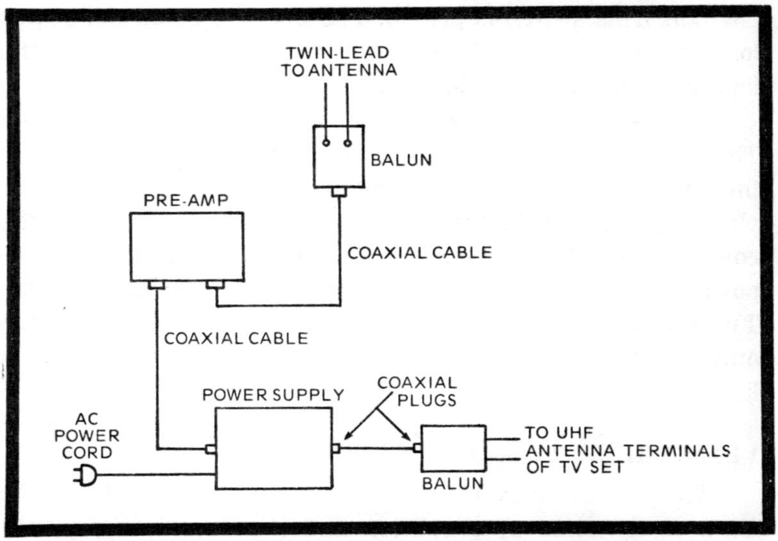

Fig. 7-6. Hook-up of mast-mounted UHF preamplifier.

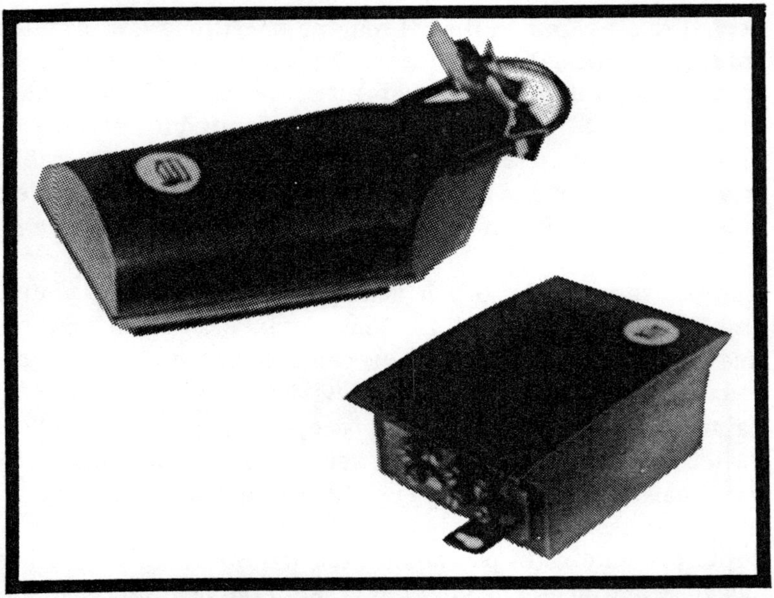

Fig. 7-7. Antenna-mounted preamplifier and power supply mounted near the receivers. (Courtesy Jerrold Electronics Corporation.)

between 76 and 100 feet long, the switch is set to extreme left position. This switch is provided so that the electrical power loss in the cable can be offset. This loss depends upon the length of the transmission line.

Figure 7-6 shows how a UHF preamplifier is connected. The twin-lead from the UHF antenna is fed to a balun. From the balun, the signal is fed through coaxial cable to the amplifier. After amplification, the signal is fed down to the coaxial cable transmission power supply and through the power supply to the balun and then to the UHF terminals of the TV receiver. As before, the electric power required by the amplifier is fed up to it through the coaxial cable transmission line.

VHF / UHF ANTENNA AMPLIFIERS

The antenna amplifiers shown in Fig. 7-7 consist of the antenna mounted preamplifier and the power supply unit which functions as a two set coupler. This antenna-amplifier

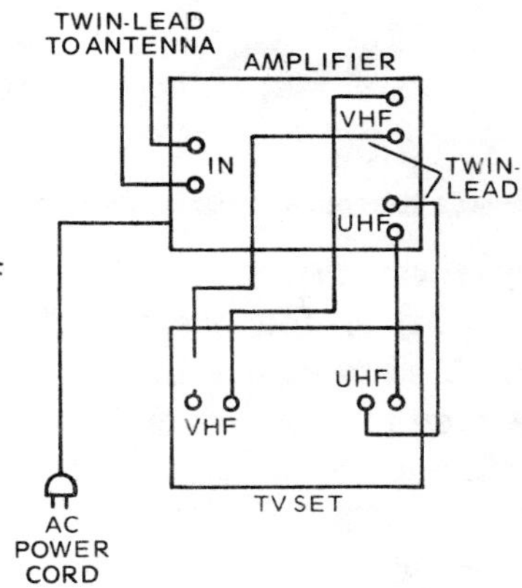

Fig. 7-8. Hook-up of VHF / UHF antenna amplifier.

system is designed for use with twin-lead cable. As can be seen, the power supply unit at the right has six screw terminals, two for the twin-lead transmission line and two each for the twin-lead cables that run to the receivers. The same manufacturer offers a similar antenna amplifier system for use with coaxial cable transmission lines. The method of connecting a VHF / UHF antenna amplifier located at the TV receiver is shown in Fig. 7-8. The twin-lead from the antenna is connected to the amplifier input terminals. The VHF output terminals are connected through twin-lead to the VHF receiver terminals and the UHF output terminals are connected through twin-lead to the UHF receiver terminals.

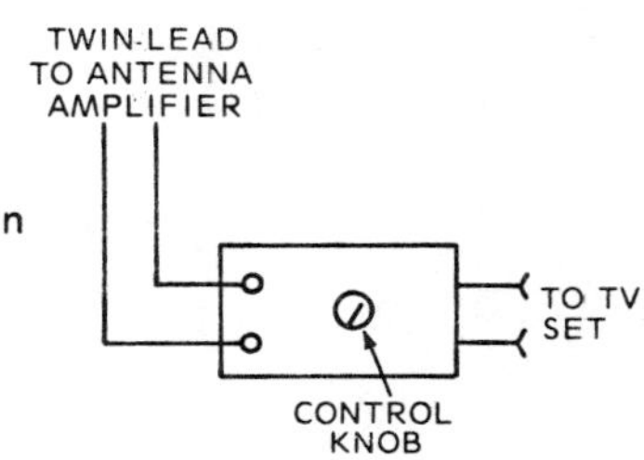

Fig. 7-9. Hook-up of attenuator in a twin-lead transmission line.

Fig. 7-10. In-line attenuator for coaxial cable.

ATTENUATORS

When an antenna amplifier is used, some channels may be amplified to such a high level as to deteriorate the quality of the picture. This problem can be set by installing a variable attenuator in series with the twin-lead transmission line as shown in Fig. 7-9. The control knob can be used to set the level of the signal fed to the receiver input. When coaxial cable is used to feed the receiver inline attenuators such as the one shown in Fig. 7-10 can be used. Figure 7-11 shows how such an attenuator is connected.

An attenuator is the opposite of an amplifier. It causes what is known as an "insertion loss."

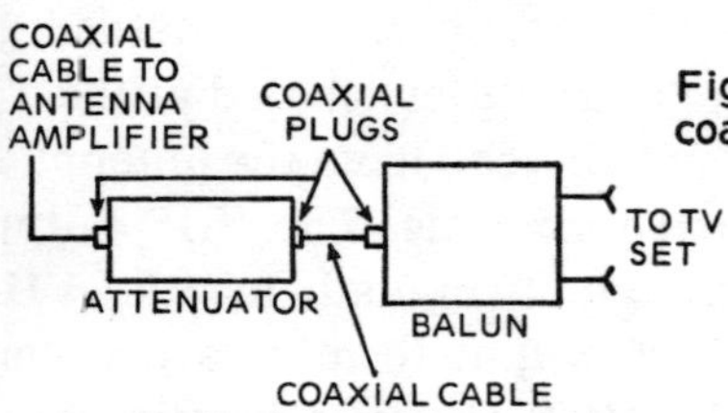

Fig. 7-11. Hook-up of in-line coaxial attenuator.

Indoor and Mobile Antennas

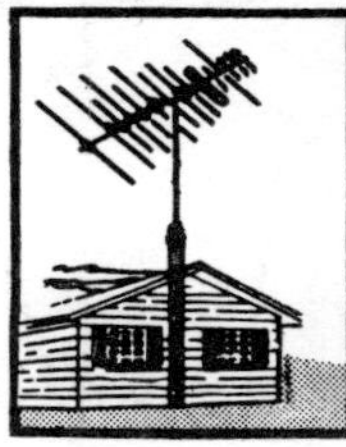

8

An outdoor antenna is always preferable to an indoor antenna. Even in strong signal areas, a well-designed outdoor antenna has better directional characteristics than indoor antennas and can provide clearer reception. When the signals are too strong, an attenuator can be used as explained in the previous chapter.

RABBIT EARS

Portable TV sets are usually provided with a built-in rabbit ear antenna or a monopole antenna. When such an antenna is not provided, a back-of-set rabbit ear assembly can be installed as shown in Fig. 8-1. A rabbit ear type of antenna is generally used only for VHF band reception and therefore its pigtails are connected to the VHF receiver terminal as shown.

UHF INDOOR ANTENNA

A wire loop about six to eight inches in diameter will often suffice as an indoor UHF antenna in strong signal areas. It is connected directly to the UHF receiver terminals as previously shown in Fig. 1-6.

A much more satisfactory UHF indoor antenna is shown in Fig. 8-2. It consists of two bow-ties and a reflector. Its directivity can be varied by simply picking up the antenna and pointing it in another direction.

VHF, UHF, AND VHF / UHF PORTABLE ANTENNAS

Typical portable antennas are shown in Fig. 8-3. The one shown in A is for VHF band reception and is essentially a rabbit ear antenna that is connected to the VHF receiver

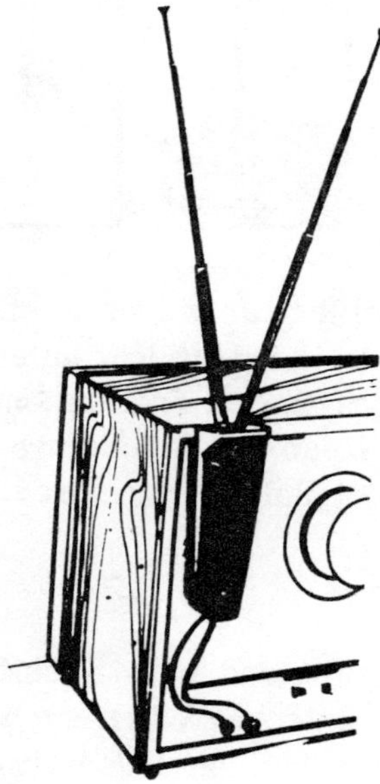

Fig. 8-1. Back-of-set rabbit ear. (Courtesy GC Electronics.)

terminals through several feet of twin-lead cable. The one shown in B is a UHF band loop antenna. This particular antenna looks like a doughnut and the width of its loop, which is made of sheet metal, improves reception throughout the UHF band. The antenna shown in C is a combination rabbit ear and loop antenna. The rabbit ear is used for VHF band reception. It has two twin-lead transmission lines, one for connection to the VHF receiver terminals and the other to the UHF receiver terminals.

POWER LINE ANTENNAS

The electric power wiring of a home can be used as a TV antenna. The device shown in Fig. 8-4 plugs into an electric power outlet. Its twin-lead cable is connected to the VHF receiver terminals.

While devices of this type are widely advertised as taking advantage of existing electric wiring for picking up television

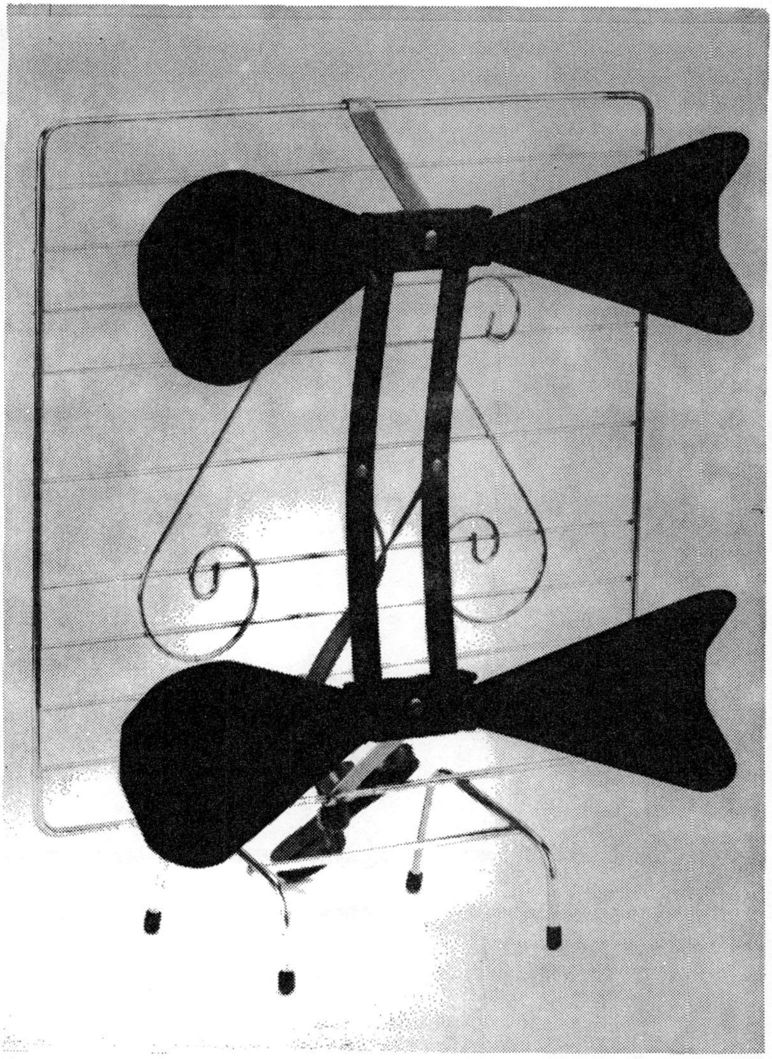

Fig. 8-2. Self-standing double-tie UHF antenna. (Courtesy Allied Radio Shack.)

signals, they are seldom as effective as other types of indoor antennas. In some cases, however, they perform surprisingly well. The main disadvantage of such an antenna is the lack of directivity which is so critical in TV reception.

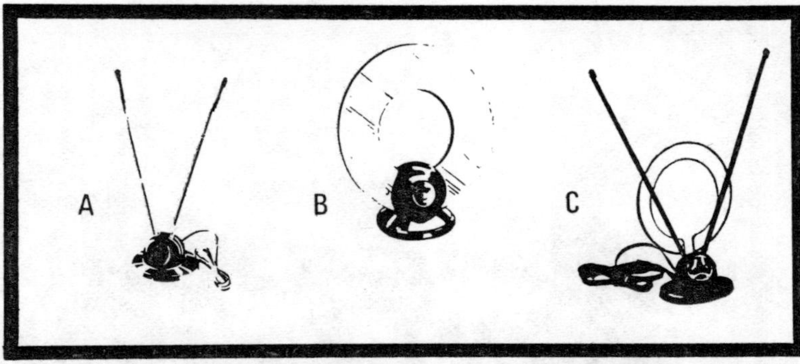

Fig. 8-3. Examples of indoor antennas. (A) VHF channels. (B) UHF channels. (C) VHF / UHF channels. (Courtesy GC Electronics.)

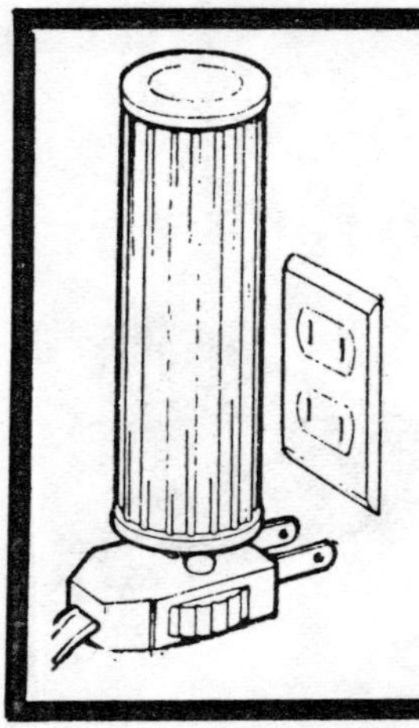

Fig. 8-4. Example of a power line antenna.

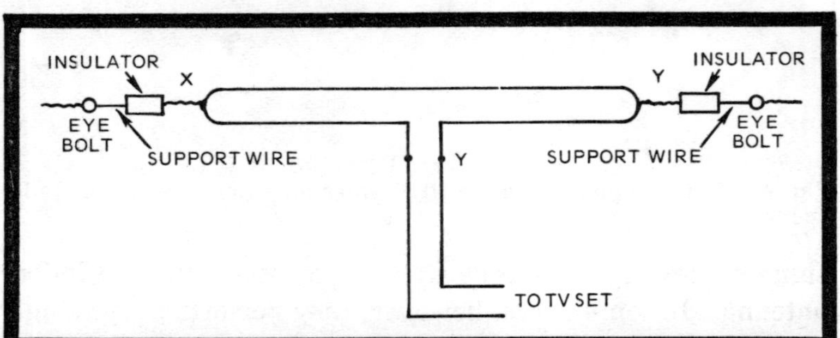

Fig. 8-5. Folded dipole antenna made of twin-lead.

HOMEMADE DIPOLES

A dipole antenna can be made of twin-lead cable as shown in Fig. 8-5. Such an antenna can be strung up in an attic, on a wall behind a TV receiver or under the carpet. It consists of a length of twin-lead from approximately 7 to 100 inches depending upon the channel to be received. Such an antenna is capable of receiving signals on more than one channel, but it is most effective at some specific channel which is determined by its length. The approximate lengths for various channels are listed in the table of Fig. 8-6.

One of the twin-lead wires is cut at the midway point Y as shown in Fig. 8-5 and another piece of twin-lead is connected to the ends where the cut was made. This other section of twin-lead is the transmission line that leads to the TV set. The far ends of the folded dipole, X and Y, are attached to a strain insulator. The other end of each strain insulator is attached to a piece of wire which is attached to an eye bolt. At points X and Y, the two wires of the twin-lead are bared of insulation and

Channel	inches	Channel	inches
2	98	9	30
3	88	10	29
4	80	11	28
5	70	12	27
6	65	13	26
7	32	14	12
8	31	68	7

Fig. 8-6. Folded-dipole length for various channels.

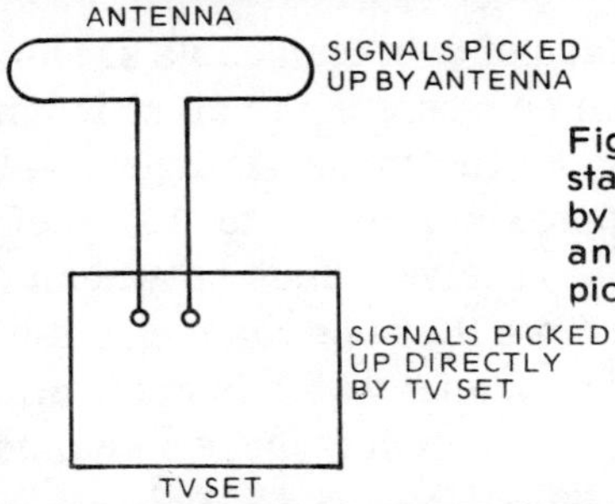

Fig. 8-7. Signals from a nearby TV station can be picked up directly by the receiver and by an indoor antenna, resulting in ghosty pictures.

A

Fig. 8-8. Recreational vehicle antenna that open and closes like an umbrella (A). In B, an example of a recreational vehicle antenna system. The antenna is raised for use or slid down for travel. The antenna shown here is the all-channel Sensar.

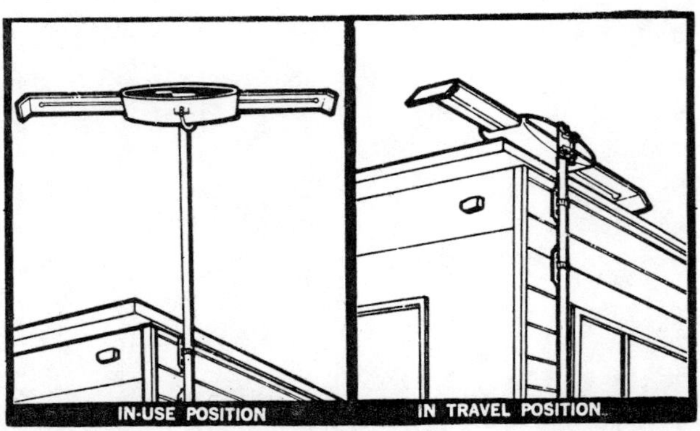

B

connected together by twisting. An alternative way of mounting such an antenna is to tack it to the eaves in the attic.

This type of antenna is bidirectional and should be mounted so that its broad side faces the direction of the station to be received.

INDOOR ANTENNA RECEPTION PROBLEMS

At locations near television stations, signals can be picked up simultaneously by the indoor antenna and by the wiring of the TV receiver itself. As shown in Fig. 8-7, TV signals are picked up by the antenna. They are also picked up directly by the TV set. Since the signals picked up by the antenna must travel to the receiver through the transmission line, they arrive later at the TV antenna terminals than the signals that are picked up directly by the internal wiring of the TV set. Since the signals do not arrive at the same time, ghosty reception is likely since two pictures appear on the screen which are not in register with each other.

VEHICLE ANTENNAS

Since so many people travel in campers and trailers, there is an increasing demand for antennas that can be mounted on recreational vehicles.

The antenna shown in Fig. 8-8A opens and closes like an umbrella. Therefore, it can be closed when the vehicle is in

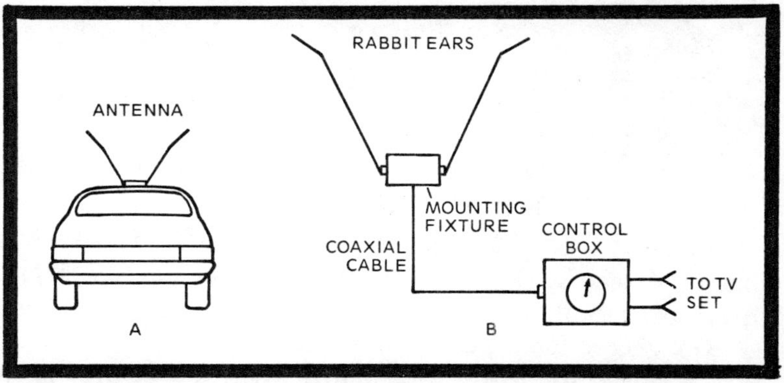

Fig. 8-9. Auto antenna installation. (A) Rear view of vehicle. (B) Electrical hook-up.

Fig. 8-10. Flat-count TV antenna for use on boats or recreational vehicles. (Courtesy Bulow International Audio.)

motion. This type of antenna is omnidirectional. This means that it will pick up signals from all directions equally well and therefore, reception is likely to be ghosty at some locations.

The recreational vehicle antenna system shown in Fig. 8-8B is raised for use or slid down when the vehicle is in motion. This particular antenna is designed for intercepting

Fig. 8-11. Mast mount TV antenna for boats or recreational vehicles. Courtesy Bulow International Audio.

both VHF and UHF band signals and is provided with a VHF / UHF splitter at the end of its coaxial transmission line.

Numerous luxury vehicles, such as limousines, are equipped with a rabbit ear type antenna on the roof. The rabbit ears are not straight but are bent as shown in Fig. 8-9. A coaxial cable transmission line is used because it is less likely to pick up interference generated by the ignition system of the vehicle. At the receiver end of the coaxial cable, there is a control box which is adjusted for best reception on the channel selected. Since the antenna is not rotatable and because the vehicle may be traveling in varying directions, picture reproduction can vary from poor to excellent. Similar antennas are also available for use on boats.

Another type of antenna for installation on a boat or recreational vehicle, the Vorta Mark II, is shown in Fig. 8-10. It has a flat mounting base and its components are sealed in a rugged plastic housing. The antenna is only 8 inches high and 18½ inches in diameter and is furnished with 20 feet of twin-lead. Another version of the Vorta VHF / UHF / FM antenna is the Mark III shown in Fig. 8-11. This model is mounted on top of a 3 to 4 foot length of 1¼ inch aluminum pipe. A chrome plated deck flange or a side mount is used for mounting this antenna. Both models are omnidirectional.

Antenna and Installation Hints

9

There are so many possible antenna system configurations that it is impossible to cover all of them in a single book. Because of the varying conditions at different location, it is impossible for the author to make specific recommendations for all situations.

ANTENNA SELECTION

There are hundreds of TV antennas on the market. Many provide comparable results. The type of antenna that is best for use at any given location depends upon the terrain and the distance to the TV stations to be received. Most antennas are listed in catalogs as being capable of receiving stations from varying distances. The specifications do not state how high the antenna must be or the dB gain of the antenna. Some specifications sheets, however, do list the antenna gain in dB, the front-to-back ratio in dB, and the directivity (beamwidth) in degrees. The gain figures (in dB) are usually listed for various TV channels. In general, the higher the dB gain figure, the higher the front-to-back ratio in dB and the narrower the beamwidth in degrees, the better the antenna.

But there are situations where wider beamwidth is preferable and excessive gain can lead to stronger signals than the TV receiver requires. There are no hard and fast rules that the layman can rely upon. Only an experienced technician or engineer can accurately interpret specifications.

SINGLE CHANNEL ANTENNA

At locations where only one TV station can be received, better results can usually be obtained by using an antenna

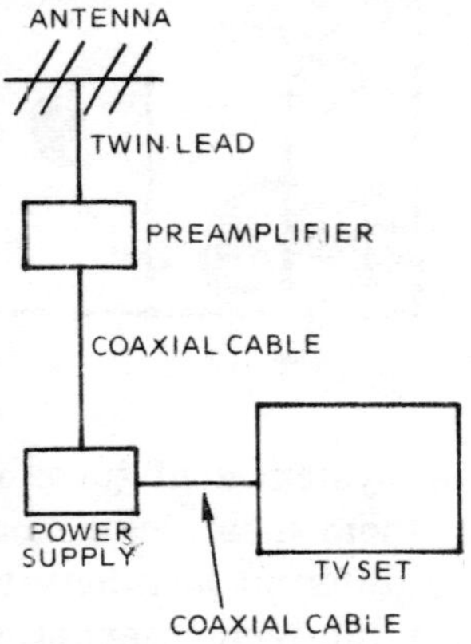

Fig. 9-1. Single-channel antenna for reception of only one UHF station or through a translator station.

designed for reception on a single channel. Such antennas are available for the VHF and UHF channels.

When the station to be received is at an extreme distance, the single channel antenna with the highest gain should be selected. In addition, a preamplifier that will pass the signals of only that channel can be installed at the antenna as shown in Fig. 9-1.

RECEPTION FROM DIFFERENT DIRECTIONS

When the stations to be received are in various directions, it makes sense to use antenna rotators since the antenna system would be less costly than providing a different antenna for each direction of reception.

An antenna system for a unique situation where the VHF stations are in one direction and the UHF stations are in another direction, is illustrated in Fig. 9-2. Here the outputs of the two antennas are fed to an antenna combiner whose output is fed through either twin-lead or coaxial cable to a VHF / UHF splitter.

LARGE CITY PROBLEMS

In a large city, such as New York, TV reception can be almost unbelievably poor in spite of the nearness to the television stations, most of which are located at one site. The TV signals are reflected by so many buildings that many reflected signals are received simultaneously, resulting in very ghosty reception.

It is for this reason that cable television systems exist in large cities such as New York. The antenna systems of the cable television system are located at a point where TV reception is excellent and the TV signals are conveyed through coaxial cables to thousands of receivers in apartment building, hotels, and homes.

Before CATV service was available in the author's apartment building, his TV receiver was connected to the MATV (master antenna television) system provided by the landlord. Because of the proximity of so many tall buildings, reception via the MATV system was unsatisfactory. He then

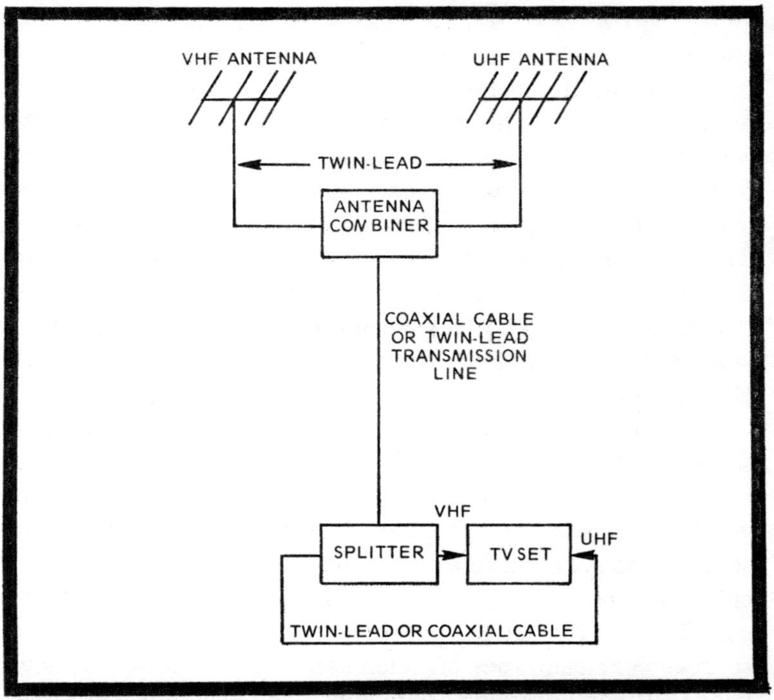

Fig. 9-2. Antenna system for use where VHF stations are in a different direction than UHF stations.

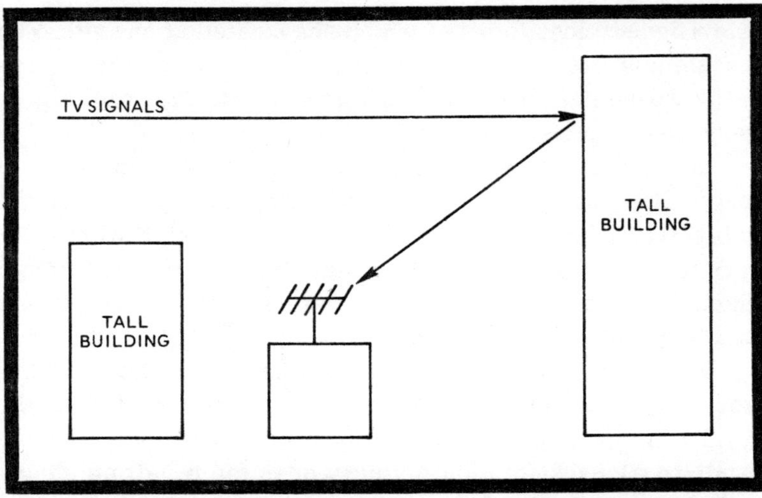

Fig. 9-3. In large cities at a location where the receiving antenna must be below the level of nearby buildings, a highly directional antenna can be used for pickup of signals reflected by a large building.

obtained a marine type rabbit ear antenna system, similar to those used on limousines, and attached it to the fence on the terrace outside of his apartment. It provided much better reception than the MATV system. One day, this outdoor antenna was accidentally disconnected from the receiver leaving it without an antenna of any kind except its own wiring. Surprisingly, the quality of color TV reception was drastically improved. This was a unique case because from the location of the TV set it was possible to look through a window and see the antennas of the TV stations on top of the Empire State Building.

Other New Yorkers living on lower floors do not enjoy the same unique situation.

At a school in midtown Manhattan, there was need for good color television reception. At the time, cable television service was not available at this location. Since the school was only ten blocks away from the Empire State Building, the signals were very strong but they came from many reflected sources. Various types of indoor and outdoor antennas were tried, but none proved satisfactory. The problem was solved however, by installing a highly directional antenna aimed at the RCA Building, in the opposite direction from the Empire State Building. Results were much better. As shown in Fig. 9-3, the signals from the Empire State Building were bounced back to the school's antenna by the RCA Building. In this particular case, the strongest reflected signal was picked up. Since the level of other reflected signals that the antenna could pick up was so low, the desiced reflected signal produced a satisfactory picture.

KING SIZE ANTENNAS

Some of the better antennas, particularly those designed for use at extreme distances from TV stations, are relatively large and heavy. They require rigid support particularly where high winds and ice loading can be expected. An example of a large VHF / FM antenna is shown in Fig. 9-4. This particular antenna is 145 inches in length. Another large antenna is shown in Fig. 9-5. This VHF / UHF / FM antenna is 132 inches long and 108 inches wide. The layman may wonder why such antennas are as large as they are and why they consist of so many elements. To be highly directional and to provide reasonably high gain at all channels, the antenna is necessarily complex and its design required unusually high engineering skills. Figure 9-6 shows the construction details of the antenna previously shown in Fig. 9-5.

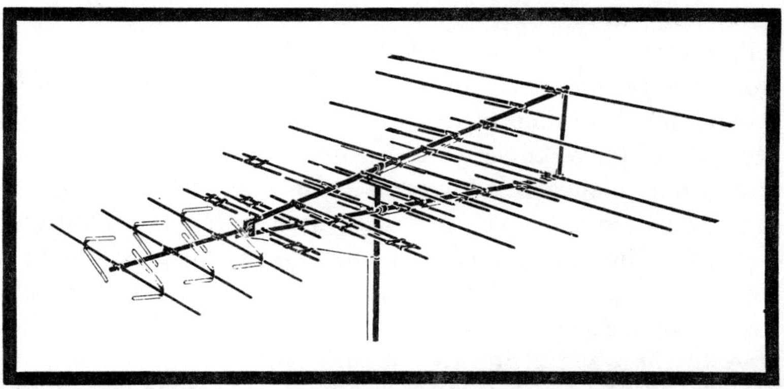

Fig. 9-4. Large VHF / FM antenna, 145" in length. (Courtesy Channel Master.)

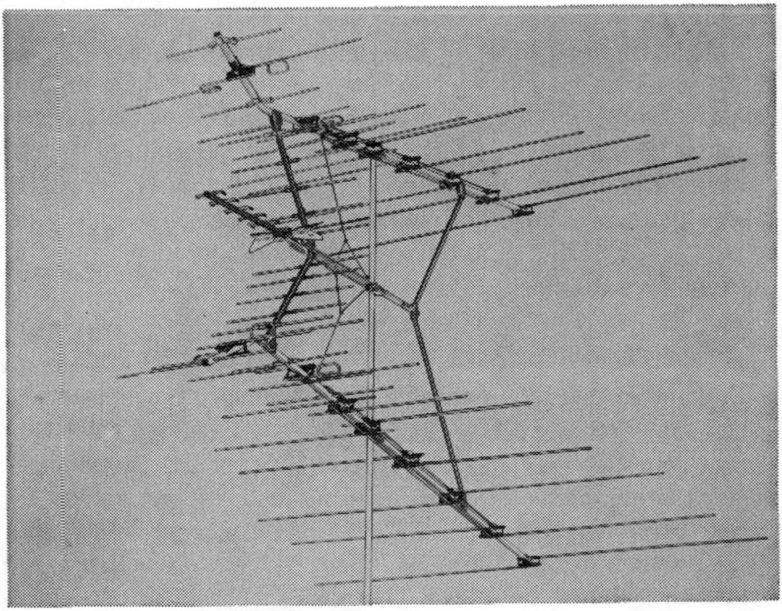

Fig. 9-5. Large VHF / UHF / FM antenna, 132" long and 108" wide. (Courtesy RCA Corporation.)

TRANSMISSION LINE CONNECTIONS

An antenna may have only one pair of output terminals or two, depending upon its design. Nearly all television antennas are designed to be connected to 300-ohm twin-lead even if coaxial cable is to be used as a transmission line. Figure 9-7 shows twin-lead connected to wing nuts of an antenna. Some other antennas are equipped with screw terminals. The twin-lead can be connected directly to the antenna terminals by baring the insulation of the two wires and wrapping each wire around a terminal. For a neater and more reliable installation, a solderless lug such as the one shown in Fig. 9-8, can be attached to each twin-lead wire with a crimping tool.

After the connections have been made, it is a good idea to apply a weatherproofing compound to the antenna terminals, as shown in Fig. 9-9. Compound should also be used for sealing the antenna end of the twin-lead.

The basic components of a guyed roof mount installation (antenna not shown) are illustrated in Fig. 9-10.

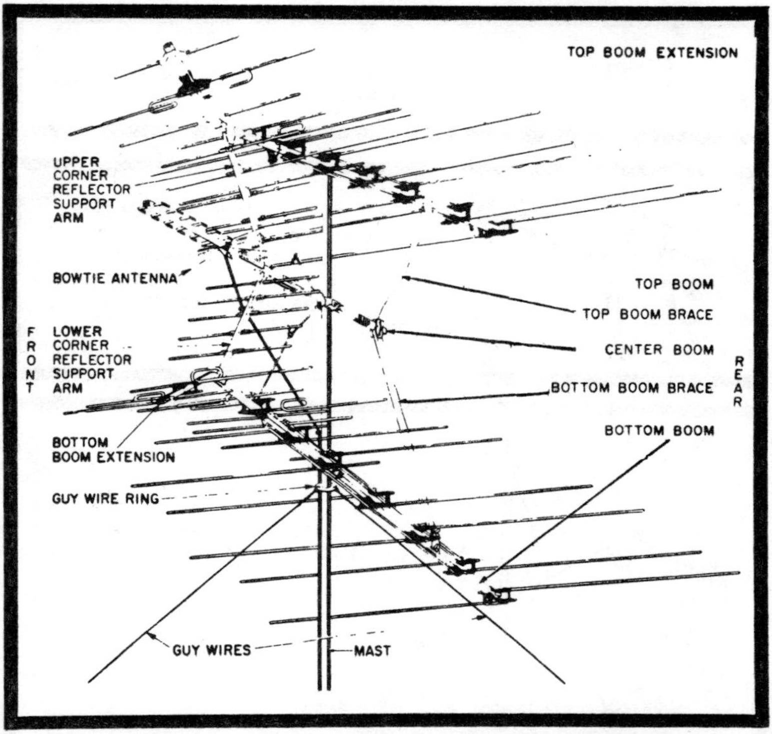

Fig. 9-6. Construction details of antenna previously shown in Fig. 9-5. (Courtesy RCA Corporation.)

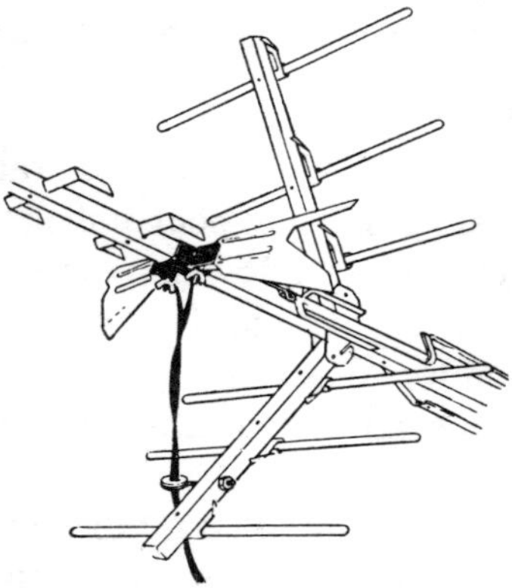

Fig. 9-7. Wing-nut twin-lead terminals. (Courtesy RCA Corporation.)

Fig. 9-8. Solderless lug for twin-lead cable.

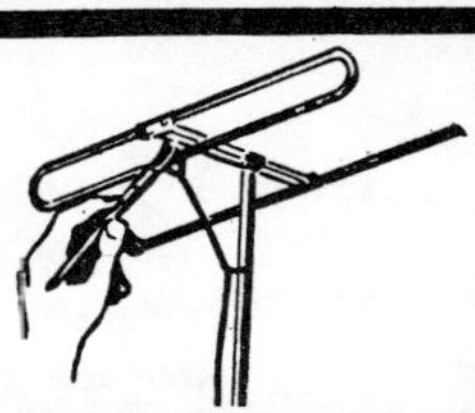

Fig. 9-9. "Liquid tape", made by GC Electronics, or other weatherproofing compound, can be applied to the output terminals of the antenna and to seal the end of the twin-lead.

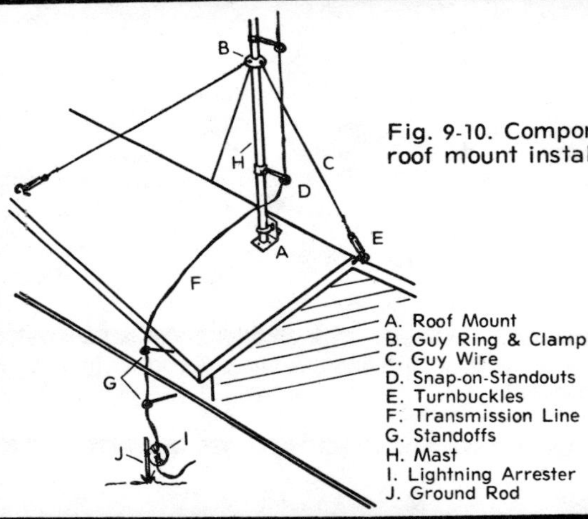

Fig. 9-10. Components of a guyed roof mount installation.

PARASITIC ANTENNA SYSTEMS

TV reception in the bottom of a valley at a great distance from TV stations may be impossible or very poor unless the receiver is connected to a CATV (cable television) system. However, under some circumstances, a "passive repeater" can be used to make TV reception possible, even if it is marginal. A passive repeater consists of two antennas connected back-to-back, as shown in Fig. 9-11.

The passive repeater should be mounted on a hilltop with one antenna aimed in the direction of the stations to be received. The other antenna should be aimed at the antenna

(C) of the TV receiver location. There should be a clear line-of-sight path between this antenna and the passive repeater. And, it may be necessary to use an antenna amplifier at the TV receiver location.

The antennas of the passive repeater (A and B) should have extremely high gain and are connected to each other through a short length of twin-lead. If the gain of each antenna is 15 dB, for example, at the desired channel frequency, the gain of the passive repeater will be 30 dB, a voltage gain of 31.62 times. Assume that the signal level at the terminals of the receiving antenna (A) is 3000 microvolts. This means that a 3000-microvolt signal will be fed to the transmitting antenna (B).

Assume that channel 13 is being received. At that frequency (213 MHz), the free-space transmission loss between the antennas is 84 dB when the passive repeater and the receiver's antenna are 1 mile apart. At ½ mile, the transmission loss is 78 dB at ½ mile, and 66 dB at ⅛ mile. If the repeater and the receiver's antenna are 660 feet (⅛ mile) apart, the free-space transmission loss will be 66 dB. If the gain of the receiver's antenna is 15 dB (same as that of the repeater antennas), the net transmission loss will be 36 dB (66 dB space loss minus 34 dB, the sum of the gains of the antennas B and C). Since a loss of 36 dB represents a 63.24 to 1 voltage

Fig. 9-11. A passive television repeater.

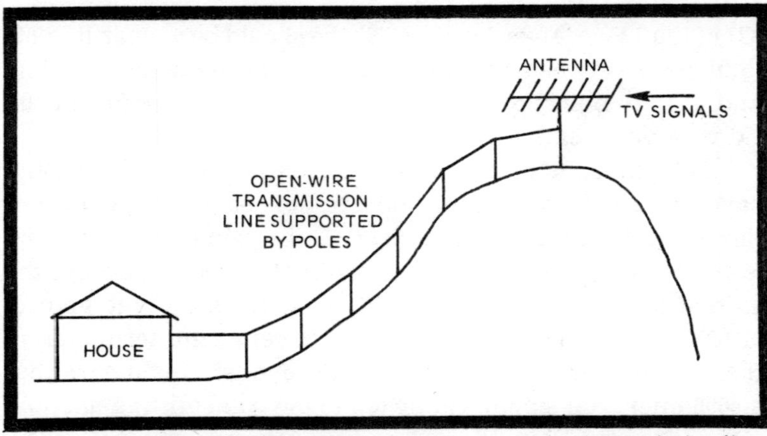

Fig. 9-12. Remotely located antenna with an open-wire transmission line.

loss, the signal level at the TV receiver input will be approximately 95 microvolts—enough to produce a TV picture, but not a good one.

By adding an antenna amplifier at antenna C to provide 20 dB of "insertion gain," the signal level at the receiver input will be raised to 950 microvolts—a very adequate signal.

If the distance between antennas B and C is doubled to 1320 feet (¼ of a mile), the signal level at the receiver will drop to 475 microvolts. And if the distance is increased to 2640 feet (½ of a mile), the signal level will drop another 628 to 237 microvolts. With 1-mile spacing, the receiver input signal level will fall to about 118 microvolts.

Again assuming that antennas B and C are 660 feet apart, the receiver input signal level will be 475 microvolts if the signal level at the output of antenna A is only 1500 microvolts. And conversely, if antenna A delivers 6000 microvolts to antenna B, the receiver input signal level will be 1900 microvolts.

REMOTE ANTENNAS

Another solution to the deep valley poor TV reception problem is to install the antenna on a hilltop and connect it to the TV receiver through a long 300-ohm open wire transmission line, as shown in Fig. 9-12. The loss introduced by this kind of transmission line is only 0.5 dB per 100 feet (during dry weather) at channel 13. Thus, a 1200-foot line would cause a loss of only 6 dB, a signal voltage drop of 50%. Of course, this type of installation is only possible when the antenna is located on the TV receiver owner's property or when easement rights can be obtained.

When the use of poles to support the open wire transmission line can't be tolerated for esthetic reasons, buried coaxial cable can be used instead. A balun is required at each end. Coaxial cable, designed for direct burial is available from CATV supply firms. The losses per 100 feet of such cables, at channel 13, are as follows: 0.412-inch, 1.6 dB; 0.500-inch, 1.3 dB; 0.750-inch, 0.9 dB. These figures are for solid aluminum sheath coaxial cable. The losses per 100 feet of solid copper aluminum sheath coaxial cable are 1.5 dB, 1.2 dB, and 0.85 dB, respectively.

A 1000-foot length of 0.750-inch solid aluminum sheath coaxial cable would introduce a loss of 9 dB at channel 13. This would cause the signal level at the receiver to be approximately 1 / 3 of the level at the antenna terminals. If the resulting receiver input signal level is too low, an antenna amplifier can be installed at the antenna end of the coaxial cable to provide up to 20 dB of insertion gain. The electric power required by the antenna amplifier can be fed through the coaxial cable from a power supply at the receiver location without interfering with the TV signals.

Coaxial cable of this type requires the use of special connectors which are available from CATV equipment supply firms.

The buried coaxial cable concept is by no means novel. People who don't want a TV antenna installed on the roof of the house, often install the antenna on the roof of a garage or on top or a tower and run buried coaxial cable to the house.

GUIDE LINES FOR TV ANTENNA INSTALLATIONS

1. Select the type of antenna(s) required for your location as recommended by your TV service technician or a neighbor who enjoys good TV reception. (Not necessarily the same make and model.)

2. Assemble the antenna as per instructions furnished.

3. Install the antenna support so it will not be blown down by the wind or pushed down by ice accumulation on the antenna.

4. Wear sneakers when walking on the roof.

5. When running bolts into the roof, apply roofing compound to prevent leaks.

6. When using twin-lead transmission line, support it with 3½" standoffs; use 7½" standoffs near metal roof gutters and when the building has aluminum siding.

7. Twist the twin-lead at 2-foot intervals.

8. In fringe areas where rain or snow is frequent, use shielded twin-lead or coaxial cable transmission line for outdoor runs in excess of 30 feet.

9. Keep twin-lead away from the mast ground wire; when using twin-lead to feed two or more receivers, do not run the twin-leads near each other.

10. Apply sealing compound on exposed antenna terminals and to seal the antenna end of twin-lead.

Reception Problems

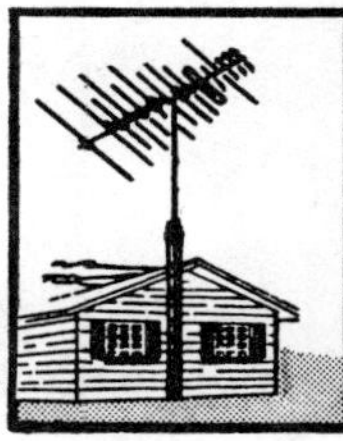

10

Television reception can sometimes be poor even when the antenna system is excellent because of other circumstances. As suggested earlier, the antenna should be located as far away from the street as possible to minimize pickup of ignition interference from passing vehicles. Reception can often be marred sporadically by interference caused by near-by radio communications transmitters.

CB RADIO INTERFERENCE

Almost one million Americans are licensed to operate approximately five million CB (Citizens Band) radio transmitters. These transmitters, when nearby, can often cause severe interference to television reception. Most often, the fault is that of the television receiver, not of the CB transmitter. A television receiver is prone to such interference because it must be designed to pass a wide band of frequencies in order to reproduce pictures, particularly color, with good fidelity. It would be possible to design television receivers which would be immune to such interference, but the cost would be considerable.

When CB interference is experienced on channels other than channel 2 only, it can often be stopped by installing a high pass filter such as the one shown in Fig. 10-1. It is connected as shown in Fig. 10-2. The twin-lead from the antenna is connected to the screw terminals of the filter and the pigtails of the filter are connected to the VHF input terminals of the TV receiver. If the interference is not sufficiently reduced by the use of one filter, two can be used as shown in Fig. 10-3. One filter is connected at the antenna terminals and the other at the receiver terminals. If the interference is still not reduced

Fig. 10-1. High-pass filter for eliminating interference from CB and amateur radio transmitters. (Courtesy Allied Radio Shack.)

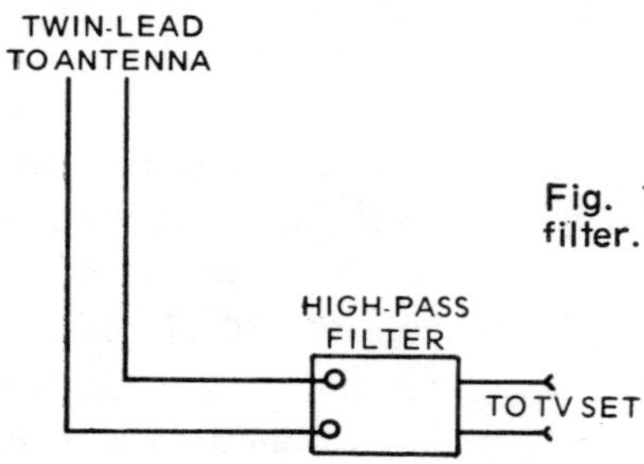

Fig. 10-2. Hook-up of high-pass filter.

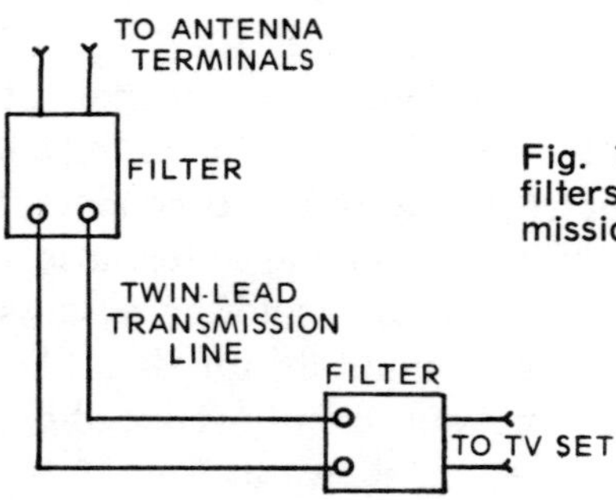

Fig. 10-3. Hook-up of high-pass filters at both ends of the transmission line.

Fig. 10-4. Hook-up of a high-pass filter when a coaxial cable transmission line is used.

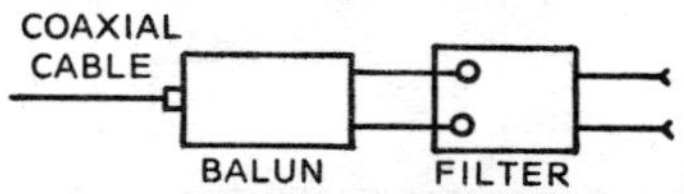

sufficiently, it may be necessary to use shielded twin-lead transmission line or coaxial cable transmission line since the unshielded twin-lead itself could be picking up the interference.

When a coaxial cable transmission line is used and this type of CB interference is experienced, the high pass filter can be connected as shown in Fig. 10-4. The coaxial cable transmission line is connected to the input of the balun, the balun's pigtails are connected to the input of the high pass filter and the pigtails of the filter are connected to the receiver terminals.

A highly effective high pass filter, such as the Miller 6167, can be connected into a coaxial cable transmission line as shown in Fig. 10-5. This particular filter has four screw ter-

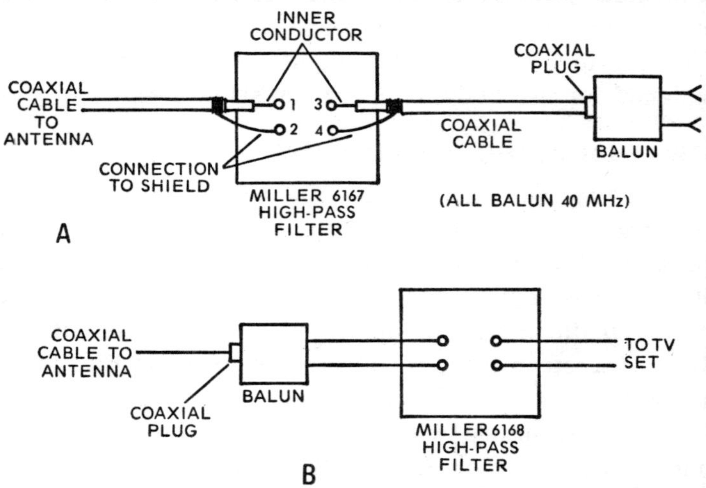

Fig. 10-5. Hook-up of high-pass filter in coaxial transmission line (A) ahead of balun. (B) after balun.

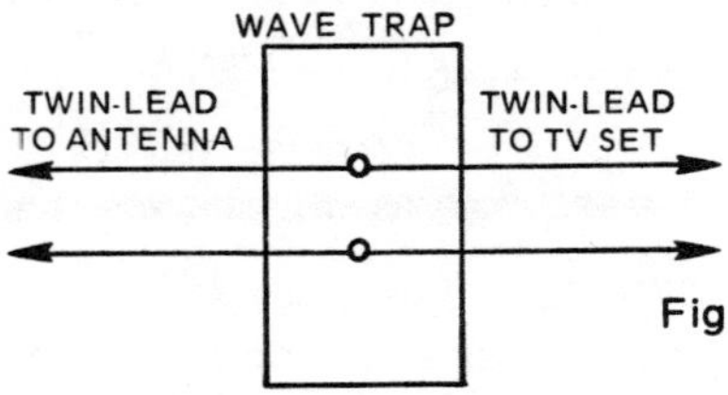

Fig. 10-6. Hook-up of wave trap.

minals, two for the input and two for the output. As shown in A, the coaxial cable is cut and the center conductor is connected to input terminal 1 and the shield of the coaxial cable is connected to input terminal number 2. The inner conductor of the coaxial cable leading to the balun and the TV receiver is connected to output terminal number 3 and the shield of that length of coaxial cable is connected to output terminal number 4. This particular filter is designed specifically for use in a 75-ohm transmission line.

An alternative is illustrated in Fig. 1-5B. Here the coaxial cable from the antenna is fed to a balun whose pigtails are connected to input terminal 1 and 2 of a Miller 6168 high-pass filter. The output terminals 3 and 4 are connected through twin-lead to the TV receiver input. This particular filter is designed for use in 300-ohm transmission lines.

AMATEUR RADIO INTERFERENCE

A high-pass filter will usually eliminate or reduce interference caused by amateur radio stations (hams) operating at frequencies below 30 MHz. To eliminate or reduce interference from amateur radio stations and other radio communications stations operating at frequencies above 30 MHz, a wavetrap can be connected as shown in Fig. 1-6. The Miller 6160 series of wave traps are contained within an aluminum enclosure with 2 terminals. The wave trap is connected as shown in Fig. 10-6 when twin-lead is used as the transmission line. At the top of the wave trap is a hole through which a screw driver can be adjusted to tune the wave trap. When the interference is being experienced, the wave trap is adjusted until the interference is greatly reduced or eliminated. The band in which the interfering station operates must be known and the appropriate type of wave trap must be selected accordingly. The Miller 6165 wave trap should be used when interference is experienced from amateur radio transmitters operating in the 50-54 MHz (6-meter) band or from commercial and public safety radio communications stations operating in the 40-50 MHz band. The Miller 6164 wave trap should be used when the interference is caused by an amateur radio station operating in the 144-148 MHz (2-meter) band or by an aviation radio service station operating in the 118-136 MHz band. A Miller 6163 wave trap should be used when the interfering station operates in the 150-174 MHz commercial and marine radio band or the 220-225 MHz amateur band.

To eliminate or at least reduce interference from CB radio stations or from amateur stations operating in the 10-meter band, the Miller 6166 wave trap should be used.

When a coaxial cable transmission line is used, the wave trap can be connected directly across the VHF receiver antenna terminals.

TRANSMITTER CAUSED INTERFERENCE

When CB radio interference is experienced only on television channel 2, it is indicated that the CB transmitter is at fault. If the interfering station can be located (usually very nearby), the owner should be requested to install a low-pass filter connected as shown in Fig. 10-7. This filter reduces the intensity of the interfering signal radiated by the CB transmitter antenna. The same technique can be used to minimize

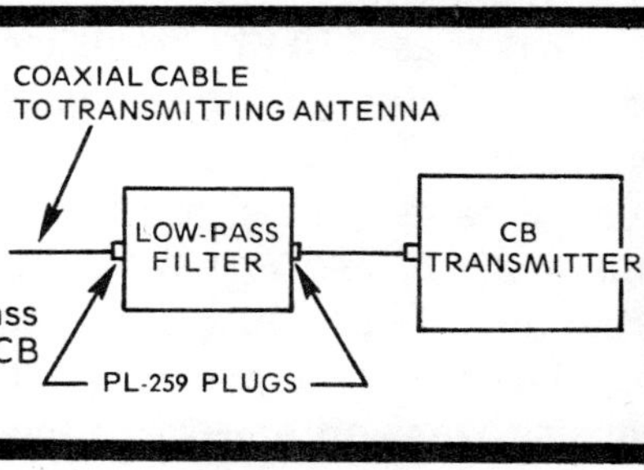

Fig. 10-7. Connection of low-pass filter to interference-causing CB transmitter.

interference from amateur radio transmitters operating at frequencies below 30 MHz.

WHO'S FAULT IS IT

The neighbors of a CBer or ham are often irate when their TV reception is marred by radio communication transmitters. They usually blame the CBer or ham even if it is the fault of the affected TV receivers. Relief is often sought by calling or writing the Federal Communications Commission to investigate and to stop the source of the interference. This seldom helps because the FCC is under-staffed and can't devote much time to helping TV set owners with their reception problems. It is better to resolve the problem personally through mutual cooperation.

Interference by CB and ham radio transmitters is not only caused to TV reception but also affects radios and hi-fi systems. Very often the interference occurs because of shortcomings in the design of the affected equipment. Of course, the interference would not occur if the CBer or ham did not use his or her transmitter. But, if appropriately licensed by the FCC, the CBer or ham has the legal right to operate a radio transmitter which conforms with FCC technical standards. And, the affected TV set owners cannot legally stop such radio transmissions.

Interference to radio reception or hi-fi listening occurs when the signal from the nearby communications transmitter is rectified (detected) by the transistors of the radio or amplifier. The remedy is to have a competent electronics technician install wave traps inside of the affected radios and hi-fi amplifiers.

In the case of TVI (television interference) caused by a CB or ham radio station, the remedies are to install wave traps or high-pass filters at the TV receiver antenna terminals, use coaxial cable instead of twin-lead as the TV antenna transmission line, and move the TV antenna as far as possible from the CB or ham station antenna. The TV antenna should be either higher or lower than the antenna of the interference-causing transmitter—not at the same height.

When the TV picture is not affected, but the voice of the CBer or ham is heard through the loudspeaker of the TV set, it is quite possible that the interference is not being picked up by the TV antenna or its transmission line. This can be determined by disconnecting the TV antenna. If the voice interference continues, the interfering radio signal is being picked up by the TV receiver wiring and is being rectified by the audio frequency (sound) amplifier within the TV set. The remedy is to have a technician install a wave trap within the TV receiver.

When a CB or ham station interferes with FM radio reception, disconnecting the antenna will determine if the interference is being picked up by the antenna system or by the wiring of the FM receiver. If the interference exists only with the antenna connected, one or more wave traps should be connected at the receiver antenna terminals. But, if the interference persists, even with the antenna disconnected, wave traps should be installed inside the receiver.

POWER LINE INTERFERENCE

If a high pass filter or a wave trap at the receiver antenna terminals does not eliminate interference from communications radio transmitters, there is the possibility that the interference is conveyed into the television or radio receiver through the electric power line. In that case, a power line filter should be installed between the TV set and the electric power outlet as shown in Fig. 10-8. A Miller 7178 filter,

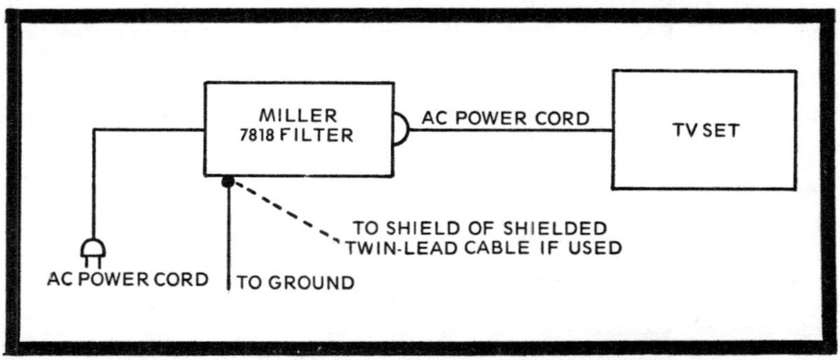

Fig. 10-8. Hook-up of an ac power line filter.

or equal, should be used. This type of a filter has a ground terminal which should be connected through the shortest possible wire to a ground rod or waterpipe.

ANTENNA SYSTEM TROUBLE SHOOTING

Table 10-1 is an antenna system trouble shooting chart. The symptoms and their various probable causes are listed as well as what should be checked and what remedies can be tried.

Since an antenna, its transmission line and other components that are exposed to the elements, are subject to deterioration, they should be replaced when television reception deteriorates.

Table 10-1. Antenna System Trouble Shooting

Symptom	probable cause	how and what to check	remedy
ghosty reception	1. Antenna incorrectly pointed.	1. Rotate antenna	1. Lock antenna into position where ghosts disappear
	2. Inadequate antenna	2. Neighbor's reception	2. Replace antenna
	3. Pickup of signals by twin-lead	3. Twist twin-lead and observe picture	3A. If twisting twin-lead clears up, leave it twisted (preferably at 1- to 1-foot intervals). 3B. Replace twin-lead with shielded twin-lead or coaxial cable
	4. Twin-lead mismatch	Move hand along twin-lead and observe picture	4. If picture is affected, shorten twin-lead 6 inches at a time until ghosts appear
snowy pictures	1. Inadequate antenna	Neighbor's reception	1. Replace antenna
	2. Inadequate or worn out twin-lead	Neighbor's reception	2. Replace twin-lead with better type
Picture flickers (especially when windy)	1. Loose antenna element	1. Antenna elements	1A. Re-secure loose element 1B. Replace antenna

Table 10-1 con't.

	2. Loose twin-lead connection	2. Check connections	2. Re-secure connections
	3. Faulty connection at coaxial cable plug	3. Check plugs	3. Cut off cable near plug and re-attach plug
Occasional streaks in picture.	1. Interference from radio communications transmitter	1. Observe picture on another TV set.	1. Install high-pass filter at VHF receiver terminals
	2. Interference from electrical devices	2A. Same as above	2A. Replace twin-lead with shielded twin-lead or coaxial cable
		2B. Turn electric devices on and off while observing picture	2B. Install noise filter at device causing interference.
Poor color pictures but black and white pictures sharp	1. Inadequate antenna	1. Neighbor's reception	1A. Replace antenna
			1B. Install antenna preamplifier

Appendix A Federal Antenna Height Regulations

Notify the Federal Aviation Administration on Form FAA 7460-1, with certain limited exceptions as set forth in Part 77 of FAA rules, of any of the following construction or alteration of an antenna structure:

(1) Construction of any new antenna structure, or alteration of any existing antenna structure, which would result in the top of the antenna or the antenna structure exceeding a height of 200 feet above the ground level at the antenna site.

(2) Construction of any new antenna structure, or alteration of any existing antenna structure, which would result in the top of the antenna or the antenna structure exceeding the height of an imaginary surface extending outward and upward at one of the following slopes:

(a) 100 to 1 for a horizontal distance of 20,000 feet from the nearest point of the nearest runway of each airport with at least one runway more than 3,200 feet in length, excluding heliports and seaplane bases without specified boundaries, if that airport is either listed in the Airport Directory of the current Airman's Information Manual or is operated by a Federal military agency.

(b) 50 to 1 for a horizontal distance of 10,000 feet from the nearest point of the nearest runway of each airport with its longest runway no more than 3,200 feet in length, excluding heliports and seaplane bases without specified boundaries, if that airport is either listed in the Airport Directory or is operated by a Federal military agency.

(c) 25 to 1 for a horizontal distance of 5,000 feet from the nearest point of the nearest landing and takeoff area of each heliport listed in the Airport Directory or operated by a Federal military agency.

(3) Any construction of an antenna structure (or any alteration of an antenna structure that would increase its height) on an airport listed in the Airport Directory of the current Airman's Information Manual.

(4) When requested by the FAA, any construction or alteration that would be in an instrument approach area (defined in the FAA standards governing instrument approach procedures) and available information indicates it might exceed an obstruction standard of the FAA.

If you intend to install towers of unusual height or at locations in close proximity to aircraft landing areas, it will be to your advantage to discuss the location and height of the antenna in detail with the appropriate FAA area office before filing your application.

Appendix B Manufacturers of Antennas and Accessories

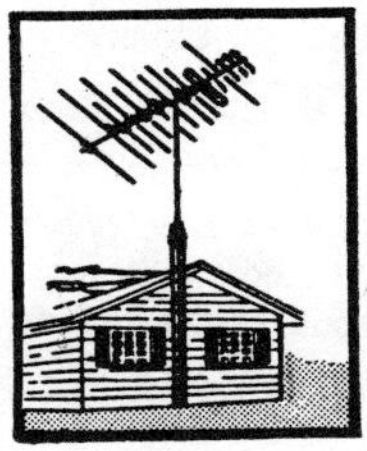

Alliance Manufacturing Company
Alliance, Ohio

Antennacraft Company
P.O. Box 1005
Burlington, Iowa

Belden Corporation
415 South Kilpatrick Ave.
Chicago, Illinois

Blonder-Tongue Laboratories
1 Jake Brown
Old Bridge, New Jersey

Bulow International Audio
225 Purchase Street
Rye, New York 10580

Channel Master Corporation
Ellenville, New York

Cornell-Dubilier
118 East Jones
Fuquay-Varina, N.C.

Finney Company
34 West Interstate St.
Bedford, Ohio

GC Electronics
400 South Wyman Street
Rockford, Ill.

Gold Line Connectors
Box 893
Norwalk, Conn.

Jerrold Electronics Corp.
401 Walnut St.
Philadelphia, Pa.

JFD Electronics
1462 62nd Street
Brooklyn, New York

Lafayette Radio Electronics Corp.
111 Jericho Turnpike
Syosset, New York

J. W. Miller Company
19070 Reyes Avenue
Compton, California

Montgomery Ward
619 West Chicago Avenue
Chicago, Illinois 60607

Mosley Electronics
4610 No. Lindbergh
Bridgeton, Mo.

Radio Shack
2615 West 7th Street
Forth Worth, Texas

RCA Parts & Accessories
Deptford, New Jersey

Rohn Manufacturing Company
Post Office 2000
Peoria, Illinois

S & A Electronics
202 West Florence St.
Toledo, Ohio

Sears & Roebuck
925 South Homan Ave.
Chicago, Illinois

South River Metal Products Co., Inc.
South River, New Jersey 08882

Superex Electronics
151 Ludlow Street
Yonkers, New York

Vidaire Electronics Mfg. Co
150 Buffalo Avenue
Freeport, New York

Winegard Company
3000 Kirkwood
Burlington, Iowa

Index

Index

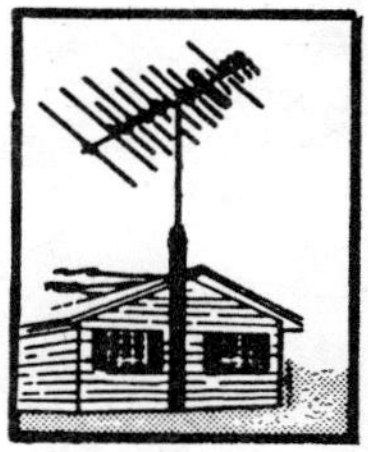

U

V

W